MICRO/NANO REPLICATION

MICRO/NANO REPLICATION
PROCESSES AND APPLICATIONS

Shinill Kang

WILEY

A JOHN WILEY & SONS, INC., PUBLICATION

Library of Congress Cataloging-in-Publication Data:

Kang, Shinill.
  Micro/nano replication : processes and applications / Shinill Kang.
    p. cm.
  Includes index.
  ISBN 978-0-470-39213-3 (cloth)
  1. Nanotechnology. 2. Manufacturing processes—Technological innovations.
3. Research, Industrial. I. Title.
T174.7.K36 2011
620'.5–dc23

                                                          2011049803

Printed in the United States of America.

10 9 8 7 6 5 4 3 2 1

To My Family

▰▰ CONTENTS

The increasing demands for micro/nanostructures or components in the field of digital display, digital imaging, data storage, optical communication, nanoenergy, and biomedicine would merit a priority in establishing the fabrication technologies for micro/nanostructures or components.

Among the various fabrication technologies for micro/nanostructures or components, the replication or molding process is regarded as one of the most suitable candidates for mass production, which may offer high quality at reasonably low cost. For this reason, researchers in both academic community and industrial sectors are beginning to actively engage themselves in pursuit of research and development in the respective field of interest.

The field of micro/nano replication or molding has recently come into existence, and thus an introductory textbook in micro/nano replication is sorely needed. A useful and desirable textbook should provide a basic (i.e., readily accessible to newcomers) interdisciplinary overview of the replicated micro/nanostructures or components, to wit: how they are designed, how the molds and stamps are designed and fabricated, how they are replicated, how the properties are predetermined and evaluated, and what are the potential uses.

The fundamental problem for both students and researchers new to the field of micro/nano replication seems to be that micro/nanopatterning and replication are rarely introduced in undergraduate-level textbooks, consequently forcing students to turn to advanced review papers, edited collections of review papers, or more advanced and specialized textbooks to begin with a learning process. Unfortunately, this body of literature is relatively impenetrable for most of the undergraduates, new graduate students, and new researchers working in the field of micro/nano replication and molding. An introductory, comprehensive and self-contained textbook would be a welcome addition for students and researchers alike interested in their respective learning field.

This book will serve as an introductory textbook on the fundamentals of micro/nano replication or molding for micro/nanocomponents. It is based on lecture notes from an introductory micro/nanofabrication course I have been teaching at the graduate school of Yonsei University in Korea over the past years. My goal is to see the textbook widely adopted and used internationally as a primary source for an introductory senior-level undergraduate or beginning graduate course covering micro/nano replication. The reader will be able to obtain a wider scope of knowledge about micro/nanomold making processes and micro/nano replication processes and the strength and weakness of each process from the book. The extensive knowledge on the micro/nanopatterning and replication processes will allow him/her to control the project on the development of micro/nanocomponents by micro/nano replication process.

However, this book is not limited to the audience in academia only, but will also be useful for the researchers and engineers in research institutes or industries. Especially in the area of micro/nano replication technology, new applications are introduced almost every day. Researchers and engineers in research institutes or industries can acquire basic and fundamental knowledge of micro/nano replication and molding.

I would like to acknowledge the financial support from the Korea National Research Foundation through the Center for Information Storage Device (CISD, an ERC directed by professor Young-Pil Park), the Center for Nanoscale Mechatronics and Manufacturing (CNMM, a 21C Frontier Program directed by Dr. Sang-Rok Lee), Nanoreplication and Micro-optics National Research Laboratory (an NRL Program), and Senior Researcher Supporting Program. Most of the research results of this book could not have been obtained without the financial and technical supports mainly from the Korea National Research Foundation for the past 10 years.

I owe a great debt of gratitude to Professor Emeritus K. K. Wang at Connell University, who was instrumental in shaping my research goal in the field of "replication."

I would like to express my warmest thanks to the professors of College of Engineering at Yonsei University for their friendship and encouragement throughout the journey of writing this book. I would also gratefully acknowledge the dedicated supports from the colleagues of CISD, CNMM, Nano Manufacturing Research Center (under the direction of professor Sang Jo Lee), The Korean Society for Technology of Plasticity (KSTP), the Korean Society for Precision Engineering (KSPE), The Korean Society of Manufacturing Technology Engineering (KSMTE) and the Korea–Japan Joint Polymer Processing Committee of KSTP and The Japan Society for Technology of Plasticity (JSTP).

Special thanks go to the former and present graduate students of the Nanoreplication and Micro-optics National Research Laboratory at Yonsei University for their invaluable assistance in preparing this book.

And last but not the least, I am very grateful to the publisher's staff, and especially Mr. Jonathan T. Rose, Editor, for their support, encouragement, and willingness to offer generous assistance during the entire book publishing project. As I cannot guarantee that this book is free of unintended errors, any corrections and suggestions from the readers are highly appreciated.

SHINILL KANG

CHAPTER 1

Introduction

1.1 INTRODUCTION

Nanotechnology is receiving more attention as an innovative technology that will lead the way to the future, along with information technology (IT) and biotechnology (BT) [1,2]. Nanotechnology permits the structure, shape, and other characteristics of a material to be controlled on a nanometer scale (10^{-9}: 1/1,000,000,000 m). Since the size of an atom or a molecule is generally on the order of 1/10 of a nanometer, nanotechnology actually provides control of the structure of a material at the atomic or molecular level (i.e., a minimal quantity of the material). While existing microtechnology is limited to product miniaturization [3–9], nanotechnology enables not only the miniaturization of components but also the creation of completely new devices based on innovative concepts since materials can be freely manipulated at the molecular level. Accordingly, nanotechnology is expected to usher in innovative changes in all industrial areas, including electronics [10–12], biology [13–15], chemistry [16,17], and energy-related fields [18,19]. Nanotechnology is regarded as being in the embryonic stage, intermediate between science and technology. However, considering the current speed of technological development and the widespread ripple effects across related industries, there is no question that a substantial market for nanotechnology will eventually be created. The National Science Foundation (NSF) of the United States predicts that the size of the nanotechnology market will exceed US\$ 1 trillion in 10–15 years [20]. A market worth US\$ 300 billion or more will be created in both the materials and the semiconductor industries, and active practical use of nanotechnology is expected in a variety of industries, including medicine, chemistry, energy, transportation, environmental science, and agriculture. For example, in the electronics industry, it is expected that new components will be developed, surpassing the limits of existing electronic devices with respect to miniaturization, speed, and power

Micro/Nano Replication: Processes and Applications, First Edition. Shinill Kang.
© 2012 John Wiley & Sons, Inc. Published 2012 by John Wiley & Sons, Inc.

consumption. The Hitachi Research Institute in Japan anticipates that the development of nanotechnology will enable the commercialization of next-generation semiconductors, in which the processing rate will be increased by a factor of 100 while power consumption is reduced by a factor of 50, together with terabyte data storage technology, in which the storage capacity will be increased by a factor of 50. Present nanotechnology market predictions pertain only to the early nanotechnology market. It is difficult to predict how nanotechnology will evolve, and what ripple effects it will create. However, considering its innovative characteristics, it is clear that nanotechnology has a tremendous capacity to create sweeping changes in the present technology paradigm.

The potential of nanotechnology has been foreseen for a long time. In 1959, the prominent American physicist Richard Feynman predicted the possibility of manipulating materials at the atomic level [21]. He anticipated that new material properties, which could not be achieved at that time, would be realized if materials could be manipulated at the atomic and molecular levels. So why has nanotechnology (which is capable of creating such an enormous ripple effect) only so recently emerged into the spotlight? The answer is, experimental results to support the theories and the development of fundamental technologies, such as the fabrication, observation, and measurement of nanoscale features, were first necessary. Furthermore, inexpensive technologies for fabricating nanostructures, which are essential to the commercialization of nanotechnology, have only recently been developed, based on existing macro- and microfabrication technologies. Among the various types of nanostructure fabrication technologies, replication-based techniques are widely used in the mass production of nanostructures due to their high repeatability, high reliability, and low cost. A replication process that can be applied to nanostructures has recently been developed, and is expected to facilitate the practical use of nanotechnology products.

Replication processes are carried out by transferring the geometry of a mold that has the negative shape of the desired product [22]. The mold is filled with material, as shown in Figure 1.1. The process generally includes heating a thermoplastic material to increase its fluidity, filling the inner geometry of the mold with pressurized molten plastic, solidifying the plastic by cooling it, and removing the resulting structure from the mold. Among the diverse materials available, thermoplastics are commonly used to fabricate replicated parts in a variety of fields because of their advantageous thermal, mechanical, optical, and electrical characteristics. However, replication processes employing thermoplastics have limited applicability to micro/nanostructured products due to material- and process-related limitations such as high processing temperature and the pressure required to fill cavities with small feature sizes. Therefore, an alternative replication process for

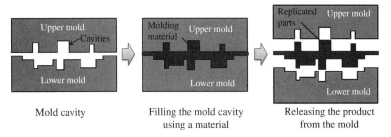

FIGURE 1.1 Concept of micro/nano replication process.

micro/nanoscale structures has been developed using thermocurable polymers or ultraviolet (UV)-curable polymers, which exist in liquid phase at room temperature and present no difficulties in filling cavities with small feature sizes. Recently, replication processes using glass or metals have also been developed, in order to overcome the limitations of polymers.

Although conventional replication processes and equipment are well established for macroscale products, specialized techniques are required for fabricating components with small feature sizes (especially nanoscale structures), such as the construction of molds with nanoscale cavities and the elimination of defects in replicated parts attributable to the large surface-area-to-volume ratios of nanoscale structures. These issues have also arisen in the replication of microstructured patterns, and a variety of modifications to conventional replication processes and equipment have already been introduced to improve the quality of the resulting microstructures. Some of the ideas employed in microreplication can be extended to nano replication technologies. However, new concepts in mold fabrication and replication methodology are also being developed to realize a greater variety of micro/nanoscale products because the modification of conventional technologies is limited to specific types of products. Since each of the various techniques for fabricating micro/nanomolds and replicating micro/nanostructures has its own pros and cons, it is necessary to select a mold fabrication method and a replication process suitable for the characteristics of the desired products. The purpose of this book is to present the techniques and principles governing specific types of micro/nano replication, as well as the characteristics of processes for fabricating micro/nanomolds, for the benefit of developers, researchers, and students interested in micro- and nanoscale products. The reader can obtain a fundamental knowledge of micro/nano replication, including the design of micro- and nanoscale components, the selection of an appropriate mold fabrication technique and replication process, the control of processing parameters, and technologies for evaluating the characteristics of micro- and nanoscale components, with application examples drawn from a variety of industries related to information storage devices, optoelectronic elements, optical communication, biosensors, and the like. Throughout this book, practical technologies will be presented for developing micro/nanoproducts via a replication process.

1.2   MICRO/NANO REPLICATION

Micro- and nanoscale products can be defined as (1) having a weight of several milligrams or less, (2) having a micro/nanoscale pattern, or (3) requiring micro/nanoscale precision [23]. Among the various techniques of fabricating micro- and nanoscale products, micro/nano replication processes are in wide commercial use because they provide high productivity and reproducibility. These procedures replicate a micro/nanoscale product or pattern, using material with appropriate optical, thermal, and/or mechanical properties, and a mold with the negative geometry of the desired product. Thermoplastic polymers are widely employed as replication materials, but thermosetting polymers, glass, metallic inks, and the like can also be used, depending on the requirements of the final products. Micro/nano replication is one of the most promising methods for fabricating nonelectronic micro/nanodevices. However, these processes recently have also been applied in fields related to the fabrication of electronic micro/nanodevices, such as integrated micro/nanooptoelectronic devices and the nanopatterning of electronic devices (nanoimprinting).

According to the type of replication material and the processing conditions, micro/nano replication techniques can be categorized into (1) micro/nanoinjection molding, (2) hot embossing (thermal imprinting), (3) UV imprinting, and (4) high-temperature micro/nano replication.

Micro/nanoinjection molding uses established injection-molding equipment to fabricate polymer micro/nanoproducts [23]. The accumulated know-how and machining technology of conventional injection molding can be applied to micro/nanoinjection molding. This type of process is especially suitable for industrialization of micro- and nanoscale components since it has the shortest cycle time among existing micro/nano replication processes. Figure 1.2 shows (a) a schematic of the micro/nanoinjection molding process and (b) pictures of mold system and molded part of nanoinjection molding. In an injection molding process, hot molten polymer is fed into a micro/nanomold cavity along a sprue, runner, and gate, as illustrated in Figure 1.2a. Since the viscosity of a thermoplastic polymer increases as it cools, and since it is difficult to fill micro/nanostructured cavities with a high-viscosity polymer, a critical issue in micro/nanoinjection molding is to overcome problems related to fluidity characteristics during the filling stage [24–26].

To overcome the fluidity problems encountered during the filling stage of micro/nanoinjection molding, a method of heating the mold and material

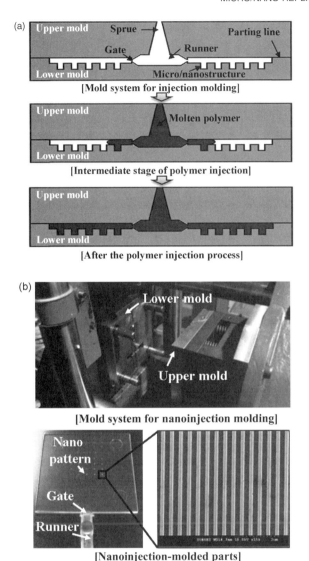

FIGURE 1.2 (a) Schematic diagram of micro/nanoinjection molding process and (b) pictures of mold system and molded part of nanoinjection molding.

above the glass transition temperature (T_g) of the material was developed, known as hot embossing (thermal imprinting) [27–28]. Figure 1.3 shows a schematic diagram of the hot embossing process. A thermoplastic material is placed on a mold with micro/nanocavities, heated (together with the mold) above its glass transition temperature, and gradually pressed into the mold. Once the micro/nanocavities are filled with the material, the mold and material are slowly cooled, and the replicated part is extracted from the

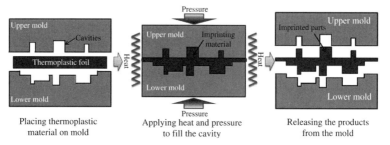

FIGURE 1.3 Schematic diagram of hot embossing process.

mold. Since the hot embossing process does not require the melting and injection of the thermoplastic material, a system for hot embossing is much simpler and cheaper than an injection molding system. However, the cycle time of a hot embossing process is much longer due to the heating and cooling time of the system. Therefore, batch processing is commonly applied to mass production of hot-embossed micro/nanostructures. Hot embossing is also suitable for the fabrication of micro/nanocomponents with high aspect ratio, which are difficult to fabricate by micro/nanoinjection molding, due to the high frication and thick solidified layer in a cavity with a high aspect ratio. Furthermore, products fabricated by hot embossing are subjected to only a small amount of residual stress since the flow is restricted to the short length within the micro/nanocavity. The characteristic of low residual stress makes hot embossing a suitable procedure for fabricating a wide range of optical components because the performance of an optical component can be degraded by birefringence, which is caused by residual stress. However, precise process control is still required to increase the size of the embossed area, which is necessary in digital display fields and other industries requiring high productivity.

Unlike the micro/nanoinjection molding and hot embossing processes, in which thermoplastic materials are melted or softened by heating in order to fill the cavities, UV imprinting uses a UV-curable resin, which exists in a liquid state at room temperature, and is polymerized by UV irradiation [29–30]. Because of the initial liquid nature of a UV-curable material, a variety of defects caused by low fluidity during micro/nano replication with a thermoplastic material can be eliminated. Therefore, UV imprinting can provide replicated parts with high-aspect-ratio micro/nanostructures and low birefringence. Moreover, the refractive index of a UV-curable resin is easily tuned, which is advantageous to the design of aberration-free imaging optics. UV-curable resins also exhibit high thermal and chemical resistance, which is important for developing highly durable products used in harsh working environments. In addition, UV imprinting can be applied to the integrating of micro/nanostructures on electronic devices, as depicted in Figure 1.4, since

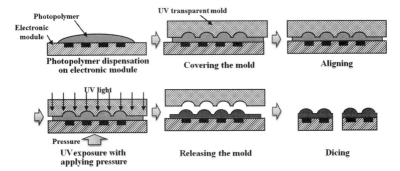

FIGURE 1.4   Schematic diagram of UV imprinting process for integrating micro/nanostructures on the electronic devices.

the process can be conducted at room temperature and low pressure. Optical alignment is also possible when a transparent mold is used.

While a typical micro/nano replication process serves to fabricate micro- and nanoscale components from a polymer material, glass or metal nanoparticle materials with a high melting point can also be used, depending on the field of application. Generally speaking, replication processes have not been applicable to glass materials, which have better optical characteristics and environmental resistance than plastic materials. However, it has recently become possible to fabricate glass products by replication due to the development of glass materials with low melting points, precisely machined mold materials with high-temperature hardness, economical high-temperature heating methods, and so on [31,32]. The glass molding method was first applied to fabricating aspherical glass lenses for a small optical imaging system. (Aspherical glass lenses possess superior optical characteristics, but cannot be economically fabricated by conventional abrasive machining techniques.) The field of application for glass molded optical parts has recently been expanded to include micro- and nanoscale components made of glass. High-temperature micro/nano replication technology combined with the use of metallic inks, which contain metallic nanoparticles, has made it possible to realize micro- and nanoscale components via a powder metallurgy process [33]. The micro/nano replication of metallic nanopowders provides a simple means of creating metallic conductive patterns, compared to the assortment of available semiconductor processes.

1.3   APPLICATION FIELDS OF MICRO/NANO REPLICATED PARTS

The development of micro/nanotechnology has been achieved through top-down fabrication methods such as E-beam lithography and other

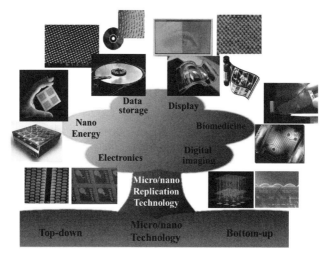

FIGURE 1.5   Micro/nano replication technology and its applications.

lithography-based techniques, and bottom-up methods such as self-assembly. Recently, micro/nanotechnology has found practical applications in a variety of fields, in combination with micro/nano replication technology. Figure 1.5 shows the areas in which micro/nano replication technology has been adopted.

1.3.1   Optical Data Storage Devices

Generally speaking, information storage devices can be classified as either magnetic information storage devices (which write and read information through changes in magnetic signals) or optical information storage devices (which write and read information through changes in optical signals). Optical information storage devices, including compact discs (CDs), digital video discs (DVDs), and Blu-ray discs (BDs), employ a variety of components fabricated via micro/nano replication processes. Figure 1.6 illustrates the application of micro/nanoreplicated components to optical information storage devices. With the increasing demands of small form factor (SFF) information storage devices for portable data storage, the miniaturization of optical data storage systems (including optical pickup units and data storage discs) is underway. Micro/nano replication has been adopted for fabricating wafer-scale optical components, in order to solve various problems that occur in the alignment of single-piece optical components.

Among the various components, the optical disc, which is the key component of an optical data storage system, is fabricated via inexpensive injection molding. This is the primary driving force behind the development

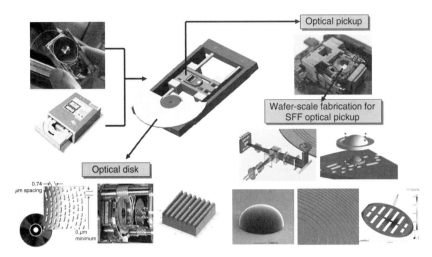

FIGURE 1.6 Application fields of components of optical information storage devices fabricated by micro/nano replication process.

of optical discs as a distributable medium. In compact disc read-only memory (CD-ROM) format, which was the first optical data storage format, information is recorded as nanopatterns on an injection-molded substrate, with a pattern width of 600 nm. The CD medium is created by forming nanopatterns on an injection-molded substrate that has been coated with any of a variety of materials. Figure 1.7 shows a schematic diagram of the cross-sectional structure of CD and BD substrates, together with atomic force microscope (AFM) measurement images of nanopatterns on an injection-molded CD substrate.

A ROM optical disc is fabricated by designing an initial nanopattern on the substrate (depending on the type of data to be stored), and the information is

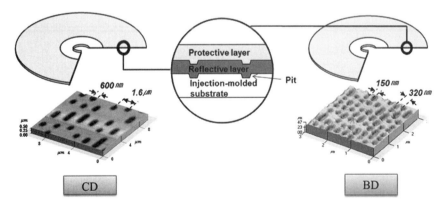

FIGURE 1.7 Schematic diagram of cross sectional structure of optical disk media and AFM measurement results of nanopatterns on injection-molded CD/BD substrate (pattern width : 600 nm and track pitch : 1.6 μm for CD, pattern width : 150 nm and track pitch : 0.32 μm for BD).

analyzed in terms of changes in the reflectivity of laser light according to the presence/absence of the pattern. In the case of a random access memory (RAM) disc, the nanopattern takes the form of tracks, which provide information about the position of data. Information is written to a RAM disc by changing the characteristics of a recording material that coats the substrate, using the heat energy of a laser, and information is read by analyzing changes in the reflectivity of laser light according to changes in these characteristics. To fabricate optical data storage medium with designated nanostructures, a series of micro/nanotechnologies are applied, including laser lithography to fabricate a master pattern, electroforming to fabricate a mold, and the final injection molding process. Figure 1.8 shows a conventional process for fabricating a metallic stamp for the injection molding of a CD or DVD substrate. A cleaned glass substrate is coated with a photoresist (PR), and laser lithography is conducted after the coated PR layer has been soft-baked. Following the development process, a seed layer for electroforming is deposited via electroless metal plating, and electroforming is carried out. The backside of the electroformed plate is polished, the metallic mold is separated from the master pattern, and a hole is punched to obtain the final metallic stamp. The metallic stamp is then installed in an injection molding machine, and an inexpensive micro/nanoinjection molding process is applied to fabricate an optical data storage disc from a substrate with a diameter of 120 mm and a thickness of 1.2 mm. Figure 1.9 shows an image of a fabricated metallic stamp for a DVD substrate, together with a fabricated DVD substrate.

The technologies for fabricating metallic stamps and optical data storage discs with nanopatterns are still commercially used for fabricating Blu-ray media, which is the most recent optical data storage format, with a storage capacity of 47 gigabytes and a data pit width of ~ 130 nm. The main

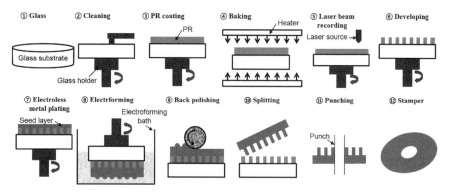

FIGURE 1.8 Schematic diagram of conventional fabrication process for metallic stamp for CD or DVD substrate.

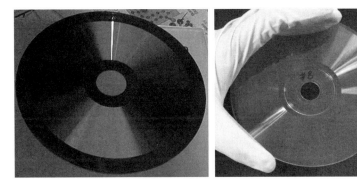

FIGURE 1.9   Images of fabricated metallic stamp for DVD substrate and injection-molded DVD substrate.

difference between the fabrication technologies for CD and BD is the mastering source: a laser for CD and an electron beam (E-beam) for BD.

The CD system is considered to be the first commercial application of nano replication technology, in which data is read from the changes in reflectance caused by nanopatterns (pattern width: 600 nm for CD) on an injection-molded substrate. The series of technologies used to fabricate micro/nanostructured devices (master patterning, mold fabrication, and replication) have also been applied in various other areas, thereby affecting the present state of micro/nano replication.

1.3.2 Display Fields

As the quality of life and visual information continue to be enhanced, the technologies related to flat-panel digital displays are rapidly developing. There are already a variety of flat-panel display systems and technologies, including organic light-emitting diodes (OLEDs), projection displays, liquid-crystal displays (LCD), plasma display panels (PDPs), and three-dimensional (3D) display technologies. Throughout the flat-panel display industry, enlargement of the display area, improvement of the image quality, and reduction of the fabrication cost are the important issues, and various micro/nanoreplicated components are playing an important role in resolving them.

LCD systems account for the greatest portion of flat display systems. Figure 1.10 shows a schematic diagram of a thin-film transistor liquid-crystal display (TFT-LCD). A TFT-LCD requires a backlight unit (BLU) to provide a surface light source for the rear side of the display since an LCD does not emit light on its own. The BLU includes a lamp, a lamp reflector, a light-guide plate, a diffuser sheet, a prism sheet, and a protector sheet. The light-guide plate is a device that creates a uniform light source by receiving incident light from the lamp, and scattering it according to a pattern formed on its surface.

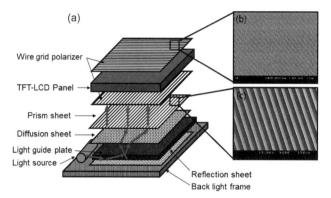

(a)

Wire grid polarizer

TFT-LCD Panel

Prism sheet

Diffusion sheet

Light guide plate

Light source

Reflection sheet

Back light frame

FIGURE 1.10 (a) Schematic diagram of TFT-LCD structures, (b) SEM image of wire grid polarizer and (c) SEM image of prism sheet.

In a typical design, the density of the scattering pattern is low on the section of the plate adjacent to the light source since a greater amount of light propagates in this section. The density of the scattering pattern is higher on sections of the plate farther from the light source, where a smaller amount of light propagates. In general, there are three methods for producing the scattering pattern. In the first method, the surface of a plastic substrate is coated with a reflective ink. In the second method, the pattern is formed via chemical or mechanical machining of a plastic substrate. In the third method, micro/nano replication is used to form appropriate micro/nanostructures on the surface of the light-guide plate, so that light is diffusely reflected by these structures. Initially, light-guide plates were fabricated via silk screening and corrosion processes, in which an organic pigment pattern was printed on a plastic substrate machined according to the intended outline, and a corrosion process was carried out using the printed pattern as a barrier. However, this process has a number of disadvantages, including increased processing time, increased costs, and degraded quality since it is divided into initial machining, printing, and material corrosion processes. Nowadays, patterned light-guide plates with controlled micro/nanostructures fabricated by micro/nano replication are widely used. Light-guide plates produced by micro/nano replication offer the advantages of reduced fabrication cost, improved productivity, and high quality. To fabricate a patterned light-guide plate, a variety of mold fabrication techniques can be employed, including superprecision machining, reflowing, etching, laser interference lithography, conventional photolithography, and electroforming, depending on the desired pattern configuration (e.g., lens, prism, or dot). Micro/nanoinjection molding is widely used for fabricating patterned light-guide plates due to its highly industry-friendly characteristics. However, hot embossing and UV imprinting processes are sometimes employed to obtain large-area, ultra-thin, or high-quality components.

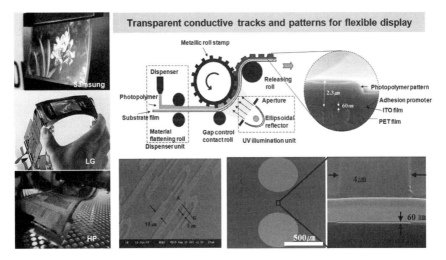

FIGURE 1.11   A schematic of UV roll nanoimprinting system to fabricate transparent conductive tracks and patterns for flexible TFT display.

Micro/nano replication technology can be applied to other types of display systems, too. Figure 1.11 shows a schematic drawing of UV roll nanoimprinting system to fabricate transparent conductive tracks and patterns on the flexible transparent substrate, which are essential for flexible TFT display. Conventionally, conductive tracks have been fabricated using metal lift-off or metal etching, where standard photolithography techniques are applied. However, since TFT should be formed on the flexible substrate for the case of flexible TFT display, roll nanoimprinting process can be an excellent candidate to realize such device. The details of the process will be explained in chapter 6.

1.3.3 Other Industries

In addition to the information storage and display applications described above, micro/nano replication is regarded as an essential technology for the practical development of micro/nanobiodevices due to the disposable nature of such devices, and relevant research is underway. Required structures for a micro/nanochannel, a micro/nanomixer, and a micro/nanoreactor for lab-on-a-chip technology have also been developed via micro/nano replication. In the optical communications field, micro/nano replication is employed to fabricate passive optical devices, such as optical waveguides, photonic crystals, microlens arrays, and optical ferrules, for the transfer, distribution, amplification, and filtering of optical signals, and the alignment of optical fibers. A series of active devices have recently been developed via the micro/nano replication process, including a vertical cavity surface-emitting laser

(VCSEL) diode and an active filter. In the semiconductor industry, nanoimprint lithography (which is based on the same principle as the nano replication process) is being applied as a next-generation patterning technology, which overcomes the limits of existing photolithography techniques. In the energy field, micro/nanoprocesses have been combined to fabricate a solar cell module coupled with a micro-Fresnel lens and a micro/nanolight-collecting device to improve the light efficiency of a solar cell, and a nanostructure electrode to enhance the reactivity of a fuel cell. In the imaging field, the technology is being applied to replicate SFF plastic/glass lenses with micro/nanoscale shape or precision, in response to the miniaturization of imaging systems. Technology based on the micro/nano replication process has been employed to fabricate a microlight-collecting device array for an image sensor, and a microlens array optical system. In particular, a wafer-scale image sensor module has been introduced, based on a concept to which the micro/nano replication and wafer-scale package processes are applied, and the development of related technologies is actively underway.

1.4   REQUIRED TECHNOLOGIES FOR MICRO/NANO REPLICATION

Since there are various micro/nanofabrication technologies, each with its own pros and cons, a person with an overall insight into these technologies (including micro/nano replication) could play a very important role in the mass production of micro/nanocomponents. As interest in micro/nanotechnology has grown, there have been a variety of technical books and research papers concerning the principles and effects of processing parameters for specific micro/nanomachining or fabrication technologies. For the development of micro/nanodevices, it is important to have the designing skills to employ whichever manufacturing process is most appropriate, taking into account the advantages and limitations of each fabrication technique, as well as the function and geometrical characteristics, marketability, and working environment of the product. For the sake of commercialization, it is especially important that one's designing skills include micro/nanocomponents fabricated via micro/nano replication since this process is most suitable for the mass production of high-quality, low-cost micro/nanocomponents. The goal of this book is to present the fundamental principles of micro/nano replication, based on years of experience in the field, necessary for students and researchers to become involved in research or technological development in this area. The reader will obtain a detailed knowledge of the micro/nanopatterning and replication processes, and the strengths and weaknesses of each process. An extensive knowledge of the micro/nanopatterning and

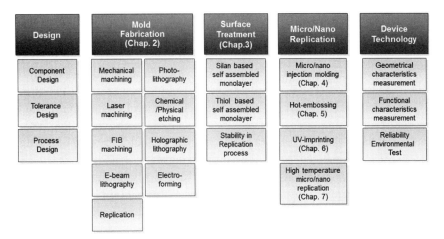

FIGURE 1.12   Overview of the steps involved in the replication process for micro/nanocomponents.

replication processes permits one to control the development of micro/nanocomponents via micro/nano replication.

To fabricate micro/nanocomponents via a replication process, it is necessary to prepare an appropriate mold with micro/nanocavities. Figure 1.12 illustrates the steps involved in the replication process for micro/nanocomponents [34]. The first stage in the fabrication of micro/nanocomponents is the design process, which consists of component design, tolerance design, and process design. In the process design step, mold fabrication, surface treatment, and micro/nano replication methods are selected to satisfy the specifications of the final component. After the mold fabrication and replication processes have been carried out, the device technology must be confirmed. This typically includes the measurement of geometrical and functional characteristics, and a reliability test.

In the design stage for micro/nanoreplicated components, component design, tolerance design, and process design should be considered. In the component design, the shape of the device is optimized via simulation or performance analysis to meet the required specifications. The component design must be conducted in conjunction with the tolerance design to control the various errors that may result from the fabrication or working conditions of the component. In the tolerance design, it is also necessary to determine a fabrication process that satisfies the tolerance requirements of the product while ensuring optimum economic efficiency and marketability. In particular, in micro/nanotechnology, it is often necessary to combine different fabrication methods to develop a new type of component, which may be difficult to accomplish with existing fabrication technologies alone. Therefore, a designer of micro/nanoreplicated components should understand not only the physical performance of a structure but also the fabrication process, working conditions, and marketing environment. It is essential to have a general understanding of the characteristics of the specific technologies used in the micro/nano replication process, as well as the characteristics of its fields of application. The first half of this book contains a brief introduction to the characteristics of each technology used in the fabrication of micro/ nanoreplicated components, while the second half discusses the design, fabrication, and testing of micro/nanocomponents, including practical examples. These examples will provide an understanding of how this technology is applied in actual industrial sites, and will assist researchers in designing processes and configurations for developing micro/nanoproducts.

Since micro/nano replication requires a mold with the negative configuration of the final product, it is essential to understand the mold fabrication technologies that can achieve the configuration and tolerance of the design. Micro/nanopatterning is widely used for fabricating micro/nanocavities, and the procedures can generally be classified as point-by-point machining techniques or lithography techniques. Mechanical machining, laser machining, focused ion beam machining, and E-beam lithography are included among the point-by-point techniques, in which a designed structure can be obtained via a combination of machining technologies, carried out to a specific point and a precise stage using tool-path data. Projection lithography, interference lithography, and the like are included among lithography techniques, in which large-area micro/nanopatterns, especially those larger than 1 mm $\times$ 1 mm, can be obtained via a single-exposure process using a photoresist. Although structures prepared via a patterning process can be used directly as molds in the micro/nano replication process, silicon, polymer, or metal molds (fabricated by silicon etching, polymer replication, or metal electroforming) are more commonly used, to ensure their durability against chemical and mechanical damage and/or releasing properties. Chapter 2 discusses the principles and characteristics of the various micro/nanofabrication processes that can be used to fabricate micro/nanomolds. In addition to the usual micro/nanofabrication technologies, the reflow process, which is widely used to fabricate microlenses for digital imaging and displays, and holographic lithography, a patterning method for creating periodic nanostructures over a large area at low cost, are discussed in detail. The electroforming process for the realization of highly durable metallic micro/nanomolds is also discussed in detail.

In conventional replication, the technique of coating a mold with a silicon-based releasing agent is widely employed to overcome the releasing problem caused by the adhesive force between a mold and a replicated product. The releasing problem is more critical in micro/nano replication because of the low rigidity and high surface-area-to-volume ratio of a micro/

nanostructure. However, it is impossible to apply existing releasing agents to micro/nanomolds since the releasing layer that is formed on the surface of the mold is thick enough to influence the micro/nanopattern. As a method for improving the releasing characteristics of a mold without influencing the micro/nanopattern, surface treatment using a self-assembled monolayer (SAM) is essential to micro/nanotechnology. Various types of materials and processes can be used to apply an SAM antiadhesion layer, depending on the type of mold and micro/nanostructures in question. Chapter 3 discusses the materials and processes for applying a SAM antiadhesion layer to a silicon/glass or metallic mold. The reliability of the SAM antiadhesion layer in micro/nano replication is also examined by analyzing the influence of an SAM on the micro/nano replication conditions.

A variety of replication materials and processes can be employed to fabricate micro/nanostructures. The developers of micro/nanoreplicated products are expected to cultivate the ability to select the most suitable replication materials and processes for the target product, based on an understanding of the characteristics of the various replication methods. Generally speaking, micro/nano replication techniques can be classified as injection molding, hot embossing, UV imprinting, or high-temperature replication. In injection molding, a molten polymer is injected into the mold cavities under high pressure, and then cooled until it solidifies. In hot embossing, both the mold and the replication material are heated to the glass transition temperature of the material, and then pressed to replicate the pattern from the mold to the material. In UV imprinting, micro/nanostructures are fabricated using a liquid-state, UV-curable resin at room temperature and relatively low pressure, and the liquid state of the initial material eliminates the fluidity problems that interfere with the filling of the micro/nanocavities. High-temperature replication is particularly applicable to the replication of metallic or ceramic materials.

Chapter 4 is intended to help readers understand the basic injection molding process, and presents a technique of efficiently heating the mold surface to improve the transcribability of the molded micro/nanostructures. A theoretical basis for the generation of the solidification layer is discussed, as well as methods of preventing this layer from forming in optical data storage media, by using an intelligent mold system. Chapter 5 is concerned with the hot embossing (or thermal imprinting) process. A process optimization technique is introduced for minimizing various defects occurring during the hot embossing process, together with an analysis of the adhesion force between the polymer material and the mold surface. Methods for measuring the geometrical and optical characteristics of micro/nanoreplicated components are also presented via a variety of examples. A design, fabrication, and evaluation technique is introduced for the optical pickup components of an SFF optical data storage system and patterned media, as a practical application of hot-embossed micro/nanostructures. Chapter 6 discusses the UV imprinting process and its applications. The photopolymerization mechanism and a method of analyzing the degree of photopolymerization using the Fourier transform infrared (FTIR) transmission spectrum are briefly described. The design and construction of a UV imprinting system and process control are also considered. The effects of processing conditions on the elimination of microair bubbles, replication quality, and residual layer thickness are examined. Finally, some practical applications are presented, such as wafer-scale integration of micro/nanocomponents in an optoelectronic circuit, eight-stepped diffractive optical elements, and roll-to-roll conductive tracks fabricated by nanoimprint lithography. In addition to the polymer-based micro/nano replication technologies discussed in the previous chapters, Chapter 7 introduces high-temperature micro/nano replication technology for metallic and glass materials. The micro/nano replication of a metal is accomplished by the sintering of metallic nanoparticles. Chapter 7 presents the technologies used for mold fabrication, surface treatment, process control, and characteristic analysis for the realization of micro-conductive tracks over a large area via high-temperature replication. Besides these, the chapter describes a replication process that can be applied to fabricate micro- and nanoscale components made of glass, which provides better optical characteristics and environmental resistance than plastic materials. In particular, a technology for fabricating a tungsten carbide (WC) mold with a micro/nanopattern for the glass molding process is concurrently discussed.

To evaluate the characteristics of micro/nanoreplicated components, both geometrical and functional characteristics should be measured and analyzed. Since the performance of micro- and nanoscale components depends on shape, it is very important to ensure their geometric accuracy. To analyze the geometrical characteristics of micro- and nanoscale components, a surface profiler, an optical microscope, a white-light interferometer, an AFM, a scanning electron microscope (SEM), and other such instruments can be used, depending on the characteristics of the materials and the shapes to be measured. However, since there is no common technique for analyzing the functional characteristics of micro/nanocomponents, specialized equipment or a specialized setup is required to measure these characteristics for any given component. Miniaturization of existing measurement systems for macroscale products has been widely employed in the measurement of micro/nanoscale components, and specialized equipment that can accurately analyze the characteristics of particular components is constantly being pursued (e.g., the near-field scanning optical microscope, or NSOM). Since the working performance of a micro/nanostructured device is just as important as the shape and characteristics of a single micro/nanopattern, a method for characterizing device performance should be developed for each application.

The latter half of this book is focused on the design, fabrication, and evaluation of micro/nanocomponents for specific applications. Chapter 8 describes the design and evaluation techniques used for UV-imprinted micro-Fresnel lenses for LED illumination systems, which are expected to occupy a large part of the market in the illumination industry. Chapter 9 describes the design and evaluation techniques used for both hot-embossed separate microlens arrays and UV-imprinted integrated microlens arrays for optical fiber coupling. Chapter 10 describes a fabrication technique for discrete track media and patterned media for next-generation hard disk systems, as well as a method for evaluating the magnetic properties. Chapter 11 describes a process optimization technique for optical data storage media. Chapter 12 describes the design and fabrication processes for nanoreplicated photonic-crystal, label-free biosensors, together with an evaluation of their biomolecule detection capabilities.

REFERENCES

1. L. Mazzola (2003) Commercializing nanotechnology. *Nature Biotechnology* 21, 1137–1143.
2. N. Mathur (2002) Beyond the silicon roadmap. *Nature* 419, 573–575.
3. J. Lim, K. Jeong, S. Kim, J. Han, J. Yoo, N. Park, and S. Kang (2005) Design and fabrication of a diffractive optical element for objective lens of small form factor optical data storage device. *Journal of Micromechanics and Microengineering* 16, 77–82.
4. J. Lee, H. Kim, Y. Kim, and S. Kang (2006) Micro thermal design of swing arm type small form factor optical pick-up system. *Microsystem Technologies* 12, 1093–1097.
5. T. A. Ameel, R. O. Warrington, R. S. Wegeng, and M. K. Drost (1997) Miniaturization technologies applied to energy systems. *Energy Conversion and Management* 38, 969–982.
6. D. Zhao, X. Hou, X. Wang, and C. Peng (2010) Miniaturization design of the antenna for wireless capsule endoscope. *4th International Conference on Bioinformatics and Biomedical Engineering (iCBBE)*, 1–4.
7. H. Weule, V. Hüntrup, and H. Tritschler (2001) Micro-cutting of steel to meet new requirements in miniaturization. *CIRP Annals: Manufacturing Technology* 50, 61–64.
8. E. Sher and I. Sher (2011) Theoretical limits of scaling-down internal combustion engines. *Chemical Engineering Science* 66, 260–267.

9. J. Hong, Y. S. Chang, and D. Kim (2011) Development of a micro thermal sensor for real-time monitoring of lubricating oil concentration. *International Journal of Refrigeration* 34, 374–382.

10. A. L. Vallett, S. Minassian, P. Kaszuba, S. Datta, J. M. Redwing, and T. S. Mayer (2010) Fabrication and characterization of axially doped silicon nanowire tunnel field-effect transistors. *Nano Letters* 10, 4813–4818.

11. J. Yao, L. Zhong, Z. Zhang, T. He, Z. Jin, P. J. Wheeler, D. Natelson, and J. M. Tour (2009) Resistive switching in nanogap systems on SiO2 substrates. *Small* 5, 2910–2915.

12. P. O. Vontobel, W. Robinett, P. J. Kuekes, D. R. Stewart, J. Straznicky, and R. Stanley Williams (2009) Writing to and reading from a nano-scale crossbar memory based on memristors. *Nanotechnology* 20, 425204.

13. T. Kim, Y.-M. Huh, S. Haam, and K. Lee (2010) Activatable nanomaterials at the forefront of biomedical sciences. *Journal of Materials Chemistry* 20, 8194–8206.

14. L. L. Chan, E. A. Lidstone, K. E. Finch, J. T. Heeres, P. J. Hergenrother, and B. T. Cunningham (2009) A method for identifying small molecule aggregators using photonic crystal biosensor microplates. *Journal of the Association for Laboratory Automation* 14, 348–359.

15. C. J. Choi, I. D. Block, B. Bole, D. Dralle, and B. T. Cunningham (2009) Photonic crystal integrated microfluidic chip for determination of kinetic reaction rate constants. *IEEE Sensors Journal* 9, 1697–1704.

16. T. Huang and X.-H. Xu (2010) Synthesis and characterization of tunable rainbow colored colloidal silver nanoparticles using single-nanoparticle plasmonic microscopy and spectroscopy. *Journal of Materials Chemistry* 20, 9867–9876.

17. G. A. Ozin and L. Cademartiri (2011). From ideas to innovation: nanochemistry as a case study. *Small* 7, 49–54.

18. M. D. Gasda, G. A. Eisman, and D. Gall (2010) Core formation by *in situ* etching of nanorod PEM fuel cell electrode. *Journal of the Electrochemical Society* 157, B113–B117.

19. I. G. Yu, Y. J. Kim, H. J. Kim, C. Lee, and W. I. Lee (2011) Size-dependent light-scattering effects of nanoporous TiO2 spheres in dye-sensitized solar cells. *Journal of Materials Chemistry* 21, 532–538.

20. B. John (2003) Leaders discuss nanotechnology market. *Astrogram* (September), 12.

21. R. Feynman (1959). There's plenty of room at the bottom. *American Physical Society Meeting*.

22. M. Heckele and W. K. Schomburg (2004) Review on micro molding of thermoplastic polymer. *Journal of Micromechanics and Microengineering* 14, R1–R14.

23. J. Giboz1 T. Copponnex, and P. Mele (2007) Microinjection molding of thermoplastic polymers: a review. *Journal of Micromechanics and Microengineering* 17, R96–R109.

24. K. Seong, S. Moon, and S. Kang (2001) An optimum design of replication process to improve optical and geometrical properties in DVD-RAM substrates. *Journal of Information Storage and Processing Systems* 3, 169–176.

25. Y. Kim, K. Seong, and S. Kang (2002) Effect of insulation layer on transcribability and birefringence distribution in optical disk substrate. *Optical Engineering* 41, 2276–2281.

26. Y. Kim, Y. Choi, and S. Kang (2005) Replication of high density optical disc using injection mold with MEMS heater. *Microsystem Technologies* 11, 464–469.

27. S. Moon and S. Kang (2002) Fabrication of polymeric microlens of hemispherical shape using micromolding. *Optical Engineering* 41, 2267–2270.

28. Y. He, J.-Z. Fu, Z.-C. and Chen (2007) Research on optimization of the hot embossing process. *Journal of Micromechanics and Microengineering* 17, 2420–2425.

29. S. Kim and S. Kang (2003) Replication qualities and optical properties of UV-moulded microlens arrays. *Journal of Physics D: Applied Physics* 36, 2451–2456.

30. P. Dannberg, L. Erdmann, A. Krehl, C. Wächter, and A. Bräuer (2000) Integration of optical interconnects and optoelectronic elements on wafer-scale. *Materials Science in Semiconductor Processing* 3, 437–441.

31. A. Y. Yi and A. Jain (2005) Compression molding of aspherical glass lenses: a combined experimental and numerical analysis. *Journal of the American Ceramic Society* 88, 579–586.

32. W. Choi, J. Lee, W. Kim, B. Min, S. Kang, and S. Lee (2004) Design and fabrication of tungsten carbide mould with micro patterns imprinted by micro lithography. *Journal of Micromechanics and Microengineering* 14, 1519–1525.

33. S. Kim, J. Kim, J. Lim, M. Choi, S. Kang, S. Lee, and H. Kim (2007) Nanoimprinting of conductive tracks using metal nanopowders. *Applied Physics Letters* 91, 143117.

34. S. Kang (2004) Replication technology for micro/nano optical components. *Japanese Journal of Applied Physics* 43, 5706–5716.

CHAPTER 2

Patterning Technology for Micro/Nanomold Fabrication

Micro/nano replication processes, including micro/nanothermal forming (compression molding and hot embossing), UV-imprinting, injection molding, and glass micromolding, are regarded as the most promising mass production techniques for micro/nanooptics, biochips, and recording media because of their high repeatability, low cost, and versatility in the choice of replication material [1–5]. To replicate micro/nanocomponents, it is necessary to prepare mold inserts that have cavities of the same shape as the desired components. Molds can be made by any of a number of established methods, including direct machining, photolithography, electron beam lithography, laser interference lithography, wet etching, dry etching, and electroforming [6–28]. This chapter presents a general overview of patterning technologies for micro/nanomold fabrication, with particular emphasis on the reflow method and laser interference lithography. Figure 2.1 shows a flowchart of pattering technology for nano/micromold fabrication.

2.1 MATERIAL REMOVAL PROCESS

In general, nano/micropatterns can be fabricated via two approaches: etching the pattern onto a bulk material or building up a small amount of material using a self-assembly method. In this section, we review material removal techniques, such as direct machining (mechanical machining and laser aberration), focused ion beam lithography, and silicon etching. Direct machining methods, including diamond turning and laser ablation with nanometer accuracy, are receiving increased attention as a means of fabricating micro/nanooptical components with arbitrary continuous surface relief profiles, such as microlenses, gratings, diffractive optical elements (DOEs), and Fresnel lenses [29–33].

Micro/Nano Replication: Processes and Applications, First Edition. Shinill Kang.
© 2012 John Wiley & Sons, Inc. Published 2012 by John Wiley & Sons, Inc.

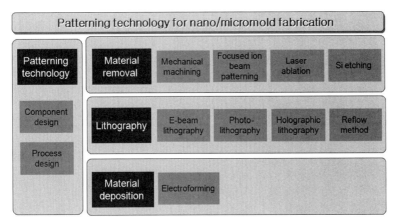

FIGURE 2.1   Flowchart of pattering technology for nano/micro mold fabrication.

2.1.1   Mechanical Machining

Mechanical machining is a technique for fabricating nano/micropatterns by using the physical energy of the tool tip to remove material. In this technology, the pattern size and shape are controlled from the tip, and the pattern shape and continuity are controlled by the moving system. However, the fabrication of a nanosized tip requires high precision and is very difficult. For this reason, mechanical machining is generally used to fabricate larger microsized patterns, such as optical elements.

Computer numerical controlled (CNC) machining and diamond turning are the current methods used for mechanical machining. If the required cavity size is larger than about 50 µm, conventional CNC machining can be used to fabricate the mold insert. However, additional surface polishing processes are often applied to shape the mold insert and improve the surface quality [30]. Microcavities with a feature size of around 10 µm and a surface roughness of around 1 µm can be obtained by diamond turning. Various mold materials can be used in this process, including metal, silicon, glass, polymer, and tungsten carbide. However, if the required feature size is smaller than about 10 mm, only materials with good machinability, such as nickel, aluminum, and copper, are used. A single diamond turning is typically used to fabricate the mold insert for micro/nanooptical components. However, if a sharp corner or small angles are required, the procedure cannot be applied since it is very difficult to exactly reproduce the desired tool shape. This process is also hampered by the slow processing time, especially for array patterns over a large area. Figure 2.2 shows a scanning electron microscopy (SEM) image of a mold insert for pyramid patterns fabricated by machining with a single diamond tool [29]. The mold material was nickel, and a master pattern with the final pyramid patterns was first fabricated by machining a V-groove

FIGURE 2.2   SEM image of mechanical machined mold inserts for pyramid patterns with a pitch of 50 μm and a depth of 24 μm.

pattern on a flat specimen, rotating the specimen by 90 degree, and then machining another V-groove pattern. The groove pitch and depth of cut were set to 50 μm and 25 μm, respectively.

2.1.2   Laser Ablation

In laser ablation, a focused laser beam is used to fabricate micro/nanostructures. When the energy of the focused beam is absorbed at the surface of the material, the surface temperature rapidly rises above its vaporization temperature, resulting in the evaporation of material. This is the key material removal mechanism of laser ablation.

However, melting due to heat conduction also occurs around the machining region, and thus sharp edges are difficult to obtain. A short pulse laser can be used to prevent this heat conduction problem. The optimum intensity of the short pulse has been theoretically analyzed to minimize melting around the machining region [35]. Material removal by laser ablation can be applied to most materials. When polymers are used, photodissociation also occurs if sufficient energy is irradiated to the samples, resulting in an increased removal rate [36]. This is why polymers are widely employed in laser ablation. The laser spot used in laser ablation has the same function as the tool tip in direct mechanical machining, and hence a precision stage is required to control the machining position. However, this procedure is time consuming. To decrease the processing time, a mask can be used, as in photolithography. A mask having contours suitable for the desired topology and a gray-scale mask with a variable transmission rate can be used to modulate the energy density distribution on the sample [37].

The image of the mask is transferred to the sample at a reduced scale via an imaging lens, such as a stepper, to obtain micro/nanopatterns with ultrahigh resolution. The laser-ablated polymer surface relief can be heat treated to obtain a smoother surface or produce a lens shape, as in the reflow method [38]. The laser-ablated polymer pattern can also be used as a mold insert for UV imprinting [39]. However, because of its low hardness, a polymer mold insert is not readily applied in injection molding and thermal forming. Additional processes, such as dry etching and electroforming, are required to transfer laser-ablated polymer patterns to mold material with a high hardness.

2.1.3 Silicon Etching Process

Etching is a removal process that utilizes an etch mask, and is accomplished by a chemical and/or physical reaction of the material. We study photolithography and E-beam lithography in Section 2.2. These processes are used to fabricate the nano/micropatterns that are employed as etch masks. There are two types of etching processes: dry etching and wet etching.

As an example, we fabricated microlenses via a wet etching process. This etching process can be employed to fabricate mold inserts for micro/nanooptical components. Depending on the shape of the component, one can use either isotropic or anisotropic etching. The substrate material, barrier shape, barrier material, and etchant are the important parameters in the etching process. The etched profile is determined by the removal rate in both the horizontal and the vertical directions, or in the crystal directions [40]. The etching selectivity, which is defined as the ratio of the removal rate of the substrate to that of the etch barrier, is a critical factor for controlling the etched profile. In Ref. [41], hemispherical dome-shaped pits were fabricated in a silicon substrate by isotropic wet etching. Figure 2.3 shows the fabrication process for a silicon mold insert for a hemispherical microlens. The etchant is based on a mixture of hydrofluoric and nitric acid solutions, with acetic acid acting as a buffer, and the etching is carried out at room temperature. The suppression of the internal residual stress of the gold masking layer is an important technique for controlling the shape and size of the hemispherical pits. The radius of the etch hole is precisely determined by one-step photolithography. The final radius of the etched hemispherical pits depends on the initial radius of the etch hole. Figure 2.4 shows SEM images of the fabricated silicon

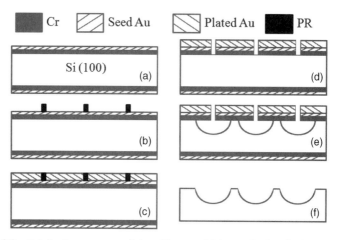

FIGURE 2.3 Fabrication process of the silicon mold insert for hemispherical microlens. Reprinted with permission from Ref. [34]. Copyright 2004, APEX/JJAP.

mold insert for a refractive microlens. The radius of curvature of the cavity was 153 μm.

Anisotropic wet etching was used to fabricate mold inserts for a V-groove optical bench and a blazed DOE [42,43]. The anisotropic etching rate of silicon in an aqueous solution of KOH yields smooth, highly accurate oriented surfaces. Figure 2.5 shows a V-groove silicon mold insert fabricated by anisotropic wet etching (20% KOH at 80°C) a silicon nitride-masked (100) silicon wafer [42]. However, the shape of the micro/nanostructure obtained from a (100) silicon wafer by this process is limited by the crystal structure of silicon. This problem can be solved by utilizing a wafer that is cut at a proper angle (111) to the plane of the silicon ingot. Using this method, a silicon mold insert can be fabricated for a blazed DOE structure with an arbitrary angle [43].

Dry etching can also be applied to fabricate a mold insert for a hemispherical microlens by controlling the process at the same etching rate in the vertical and horizontal directions [40].

Reactive ion etching (RIE) is a widely used process for fabricating microstructures such as binary optics on wafers. The procedure can be adopted to produce mold inserts for stepped DOEs and approximately aspheric refractive optics. Figure 2.6 shows the fabrication process for a mold insert for a four-stepped DOE [29]. Two masks are used to create a four-stepped structure. The first mask exposure and subsequent etching process produce a two-stepped structure with an etch depth of 5.4 μm. The second mask exposure and etching process yield a four-stepped structure. The second etch depth is 1.8 μm. Figure 2.7 shows an SEM

(a)

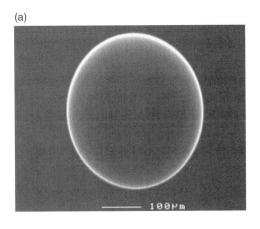

(b)

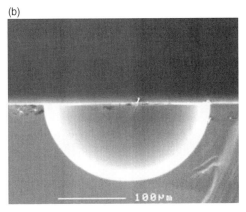

FIGURE 2.4   SEM images of fabricated silicon mold insert for hemispherical microlens: (a) top view and (b) cross section. Reprinted with permission from Ref. [34]. Copyright 2004, APEX/JJAP.

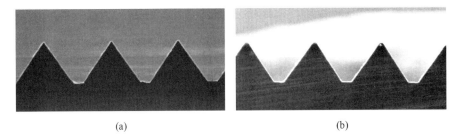

(a)   (b)

FIGURE 2.5   SEM images of V-groove pattern (a) V-groove silicon mold insert with a pitch of 250 μm fabricated by anisotropic wet etching of (100) silicon wafer, and (b) replicated polymer pattern. Reprinted with permission from Ref. [34]. Copyright 2004, APEX/JJAP.

image of a fabricated silicon mold insert for a four-stepped DOE. By patterning the barrier layer, which has a selectivity of almost 1.0 for reactive ion etching, micro/nanostructures for arbitrary-shaped optical components can be fabricated on a substrate. This transfer process can be applied to silicon, glass, and polymer substrates. The process is especially appropriate for fabricating diamond mold inserts with micro/nanostructures. Diamond offers many advantages as a mold insert material, such as high mechanical, thermal, and chemical properties [44]. The preshaped barrier pattern can be fabricated by various micro/nanomachining processes, including photolithography, E-beam lithography, X-ray lithography, photoresist reflow, and laser ablation.

2.1.4 Focused Ion Beam Patterning

A focused ion beam (FIB) is patterning process by using cutting off defined material and depositing material onto it. Generally, a gallium ion is used to mill and deposit any material. And addition of the some gas compounds makes selective etching and increases the etching velocity that is used to modify and edit nano/microdevices. Also, FIB is used to make sample for showing the cross-section view of nano/microstructure, site-specific transmission electron microscopy (TEM) sample preparation, and martial grain/film imaging.

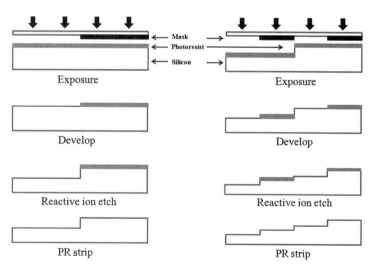

FIGURE 2.6   Fabrication process of the silicon mold insert for four-stepped DOE. Reprinted with permission from Ref. [34]. Copyright 2004, APEX/JJAP.

FIGURE 2.7   SEM image of fabricated mold inserts for four-stepped DOE. Reprinted with permission from Ref. [34]. Copyright 2004, APEX/JJAP.

2.2   LITHOGRAPHY PROCESS

In general, micro/nanopatterning is used in semiconductor fabrication. A typical lithography process consists of resist deposition, soft baking, light exposure, developing, and postbaking. The resist-deposited substrate is exposed to an electron beam and ultraviolet light. The polymer chain of the exposed electron beam is then exchanged, and resin is removed by the developing process. In this section, we study E-beam lithography and photolithography. For the fabrication of microlenses and large-area nanopatterning, we focus our attention on the reflow method and laser interference lithography.

2.2.1   Electron Beam Lithography

Electron beam (E-beam) lithography is based on the same technology as was used to develop the scanning electron microscope in 1950. It is used to fabricate nanopolymer patterns, and is similar to photolithography. The polymer chain of the exposed electron beam is exchanged, and the resin is removed by the developing process. Nowadays, this process is of primary importance in the fabrication of semiconductor large-scale integration (LSI) stepper masks [45–50].

E-beam lithography is frequently used in the fabrication of prism nanopatterns. As an example, we fabricated a pillar pattern with a diameter of 50 nm. E-beam lithography and an inductively coupled plasma (ICP) etching process were used in the fabrication of the original silicon nanomaster with nanopillar patterns. The patterned area of the silicon nanomaster was 1.5 mm × 1.5 mm. Figure 2.8 shows schematics of E-beam lithography process and SEM images

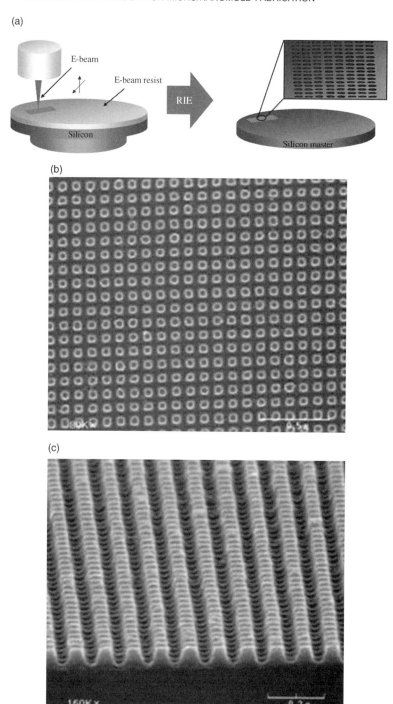

FIGURE 2.8 (a) Schematics of E-beam lithography process and SEM images of fabricated nano pillar patterns on silicon nanomaster with 50 nm pillar diameter and 100 nm pitch. (b) Top view and (c) oblique view. Reprinted with permission from Ref. [34]. Copyright 2004, APEX/JJAP

of the nanopillar array pattern with a pitch of 100 nm on the silicon nanomaster [51].

E-beam lithography can be employed to fabricate various nanopatterns using x–y stages. However, for recording media, the writing cycle must have the same diameter. For this reason, E-beam recording has been developed. As an example, we fabricated patterned media with full tracks. To fabricate the original silicon master with full tracks of nanohole patterns (hole diameter of 30 nm, pitch of 50 nm, track width of 1.3 mm, and track inner diameter of 40 mm), electron beam recording (EBR) and an ICP etching process were used. ZEP520 was spin coated on the silicon substrate as a resist for the subsequent EBR process. The coated substrate was moved to the designated position using a translation stage, and was then azimuthally rotated by the rotation stage during E-beam exposure. Following the EBR process, ICP etching was carried out to etch the silicon substrate using the electron-beam lithographed master pattern as a barrier. Figure 2.9 shows the fabrication results for the silicon master with full tracks and shows the details of the nanohole and line patterns (depth of 50 nm) [52,53].

2.2.2 Photolithography

The photolithography process employs changes in the photoresist structure caused by light exposure. Since the mask blocks the exposed light, any desired pattern can be fabricated on the substrate. Photolithography is composed of a number of processes. The photoresist is first spin coated on the substrate and the solvent is removed via the soft backing process. The substrate is then exposed to light, and exposed and unexposed photoresist are removed by the developing process. Figure 2.10 illustrates the process. Since photolithography is discussed in many other references, in this chapter we focus our attention on the reflow process. We fabricated an image sensor as an example [54–56].

2.2.3 Reflow Method

2.2.3.1 Fabrication of a Mother Lens   The reflow method, introduced by Popovic et al. [6], is the most widely used technique because of its simplicity and low cost. The procedure is carried out by first etching cylindrical patterns of photoresist onto a silicon wafer using standard photolithography. These patterns are then thermally treated to allow the surface tension to convert the photoresist patterns into lens forms. Lenses produced by the reflow method are referred to as reflow lenses, and can be fabricated at the micrometer level with acceptable surface quality. Reflow lenses can also be used as masters for glass and polymeric lenses fabricated

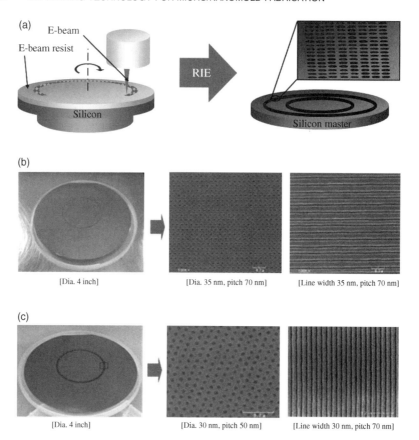

FIGURE 2.9 (a) Schematics of E-beam recording process, (b) Fabrication results for the silicon master with full tracks (line pattern with width 35 nm and pitch 70 nm), and (c) Fabrication results for the silicon master with full tracks (dot pattern with width 30 nm and pitch 70 nm). Reprinted with permission from Ref. [52]. Copyright 2009, Institute of electrical and electronics engineers.

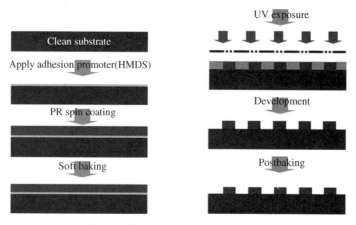

FIGURE 2.10 Photolithography processes.

by other processes, such as reactive ion etching or microreplication [57,58]. In this case, the dimensional accuracy of the final lens profile depends on the dimensional accuracy of the reflow lens master.

Many research studies have been undertaken to predict reflow lens profiles. Daly et al. [8] derived a simple equation to determine the thickness of the initial photoresist layer based on the assumptions that the reflow lens is spherical in shape and the photoresist volume does not change during heat treatment. For a more thorough mathematical description of the process, it is necessary to take several additional parameters into consideration. Sinzinger and Jahns [59] considered the effects of surface tension and gravitational energy. A quasi-spherical shape develops due to the surface tension at the resist–air interface. The weight loss of the photoresist was analyzed by Shaw et al. [60] who attributed it to volatile components and predicted a 9% loss of material during the melting process. Jay et al. [61] reported a final sag height that was 20% less than the value predicated by Daly's equation. However, little consideration was given to volume changes during the reflow process, and the derived equations failed to account for such volume changes. Furthermore, the effect of temperature and time on photoresist volume changes should also be considered.

In the present study, empirical relationships between the reflow process conditions and the volume change ratio were derived to predict the required initial volume of the photoresist patterns. To accomplish this, reflow lenses with various volumes were fabricated under different processing conditions, and the profiles of these lenses were measured. The effects of heating temperature and reflow time on the profiles of the reflow lenses were analyzed. An empirical relationship between the reflow process conditions and the volume change ratio was derived to predict the required initial volume of the photoresist patterns. To verify the empirical equation, a new microlens was designed and fabricated. The reflow lens was fabricated precisely in accordance with the empirical equation.

Photoresist islands were fabricated on a silicon substrate with a thickness of 0.5 mm and a diameter of 100 mm using a standard photolithographic processing sequence. Positive photoresist material was used with a propylene glycol monomethyl ether acetate (PGMEA) base. The substrates were first cleaned using the RCA process. After the hexamethyldisilazane (HMDS) was spin coated, the photoresist was coated and baked at 95°C. The photoresist coating was exposed using a Karl Suss MA6 at 365 nm, and developed and postbaked at 110°C. It is important that all the patterned photoresist islands are formed under the same conditions as the substrate material spin conditions, including humidity, during lithography, baking, exposure, and development. All the processes were carried out in a clean room at a constant temperature and humidity (21°C, 60%). The photoresist island had a truncated cone shape that tapered toward the edge of the resist, as illustrated in

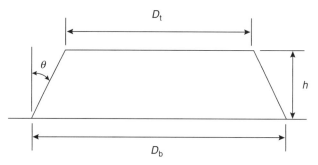

FIGURE 2.11   Shape of patterned photoresist island as truncated cone.

Figure 2.4. Because the photoresist absorbs light, the light intensity at the bottom of the resist layer is less than that at the surface, which is therefore relatively underexposed. Upon development, the resist at the top is preferentially removed, and the patterned photoresist island tapers inward in a truncated cone fashion, as shown in Figure 2.11 [62]. The volume of the photoresist island before thermal reflow is given by

$$V_{tc} = \pi \left[\frac{D_b^2}{4} h - \frac{D_b}{2} h^2 \tan \theta + \frac{h^3}{3} \tan^2 \theta \right] = \frac{\pi h}{12} \left(D_b^2 + D_b D_t + D_t^2 \right) \quad (2.1)$$

where D_b is the diameter of the bottom of the patterned photoresist island, D_t is the top diameter, and h is the thickness. The taper angle varies with exposure, development, and baking conditions. In most cases, the taper is close to $15°$. After patterning photoresist islands of various diameters and thicknesses, the top and bottom diameters, D_t and D_b and the thickness h were measured using mechanical and optical profilers. The volumes of the photoresist islands calculated using Eq. (2.1) were 475000 µm³, 52000 µm³, and 17700 µm³. The volume deviation for the same pattern size was less than 2%.

The reflow process was conducted in a controlled environment at a temperature of 21°C and a humidity of 60% in order to minimize the effects of other conditions on the shape of the reflow lens. After raising the hot plate temperature to the heating temperature T, the specimen with the photoresist island was placed on the hot plate and left there for the required reflow time t. Specimens 10 mm × 30 mm in size were cut from the patterned wafer. For larger specimens, the temperature and time were increased to allow transfer at the same heat flux. After thermal treatment, the specimens were cooled on a glass plate. To examine the influence of heating temperature and reflow time on the profiles of the reflow lenses, the reflow lenses were produced at different heating temperatures T between 130 and 220°C, and reflow times t between 50 and 200 s. Figure 2.12 shows SEM images of reflow lenses with diameters of 60, 96, and 236 µm.

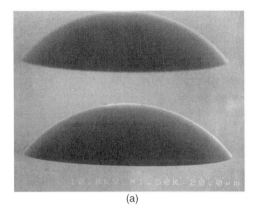

(a)

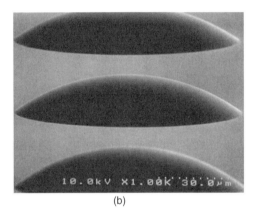

(b)

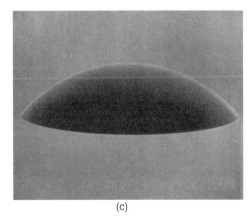

(c)

FIGURE 2.12   SEM images of reflow lenses. (a) Lens diameter of 60 μm; (b) lens diameter of 96 μm; (c) lens diameter of 236 μm.

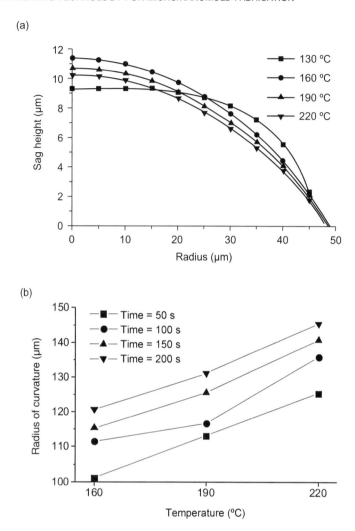

FIGURE 2.13   Effect of heating temperatures on (a) lens shape and (b) radius of curvature.

The reflow lens profiles were measured by interferometric three-dimensional profiliometry and mechanical profiliometry. Figures 2.13 and 2.14 show the profiles of lenses with a diameter of 96 μm fabricated with different reflow times and heating temperatures. The diameter of the lenses did not change during the reflow process, and the lenses produced were aspherical. Figure 2.13a shows the effect of the heating temperature on the lens profile. When the reflow time was fixed at 100s, the shape of the photoresist pattern changed as the temperature increased. At a temperature of 130°C, the surface tension was insufficient to raise the center of the photoresist island into a peak. However, at temperatures

(a)

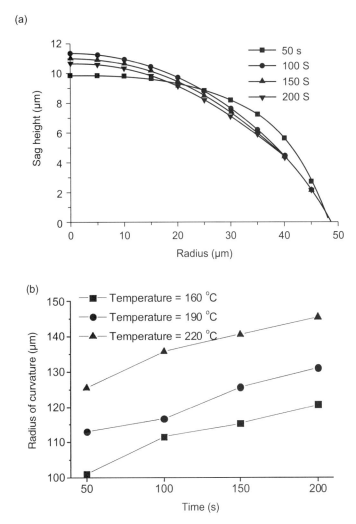

FIGURE 2.14   Effect of reflow time on (a) lens shape and (b) radius of curvature.

above 130°C, the photoresist island was converted into a convex form. Figure 2.13b shows the radii of curvature of the reflow lenses at different heating temperatures. After the photoresist island was converted into a lens shape, the radius of curvature increased and the sag height of the reflow lens decreased as the heating temperature increased. Figure 2.14 shows the effect of the reflow time on the lens profile. In Figure 2.14a, the heating temperature was fixed at 160°C. Since a hot plate is heated by conduction, the bottom and edges of the photoresist island were the first to melt. Then the edges of the island grew more rounded. As the reflow time increased, the surface tension became sufficient to raise the center of the photoresist pattern into a peak, and the

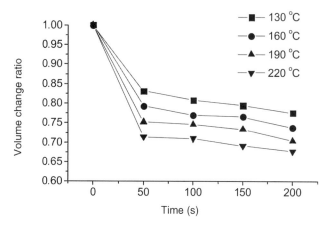

FIGURE 2.15 Effects of heating temperature on lens volume change ratio.

resist melted and formed a convex lens. Figure 2.14b shows the radii of curvature of the reflow lenses produced with different reflow times. After the photoresist island was converted into a convex lens shape, the radius of curvature increased and the sag height of the reflow lens decreased as the reflow time increased.

The lens volumes were obtained from the measured lens profiles. Figures 2.15 and 2.16 show the volume change ratio of the initial photoresist island to the final reflow lens when the diameter of the reflow lens was 96 μm. The relationship between the volume change ratio and the heating temperature is shown in Figure 2.15. Most positive photoresists have a glass transition temperature T_g between 110 and 160°C. At heating temperatures over T_g, the photoresist deforms freely, and the solvent and moisture trapped in the

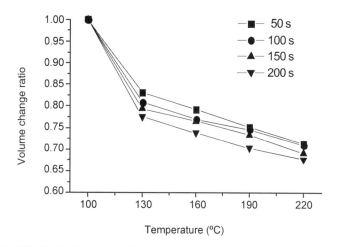

FIGURE 2.16 Effects of reflow time on lens volume change ratio.

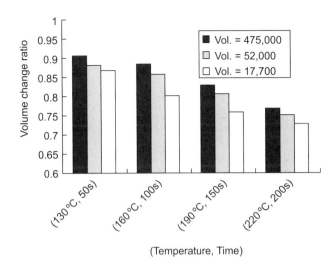

FIGURE 2.17   Volume change ratio for different volume and surface area.

photoresist are vaporized. Thus, the volume of the photoresist shrinks. The relationship between the volume change ratio and the reflow time is shown in Figure 2.16. As the reflow time increased, the volume change ratio of the photoresist decreased exponentially.

Figure 2.17 illustrates the relationship between the volume change ratio and the photoresist lens volume. At a low temperature and reflow process time, the volume change ratio is reduced as the volume of the reflow lens increases due to a lack of heat flux in the larger photoresist volumes. In order to evaporate solvent and moisture at proportional rates, the heating temperature and reflow time should be increased for larger photoresists.

2.2.3.2 Empirical Equation for the Volume Change Ratio of a Reflow Lens   The experimental reflow results demonstrated that the volume change ratio of the initial photoresist island to the final reflow lens is a function of the heating temperature T and the reflow time t, expressed as

$$\frac{V_{\text{lens}}}{V_{\text{tc}}} = f(T, t) \tag{2.2}$$

The volume change ratio decreases with T and t. On the basis of the experimental results, an exponential function was chosen to describe the volume change ratio as follows:

$$f(T, t) = \exp\left(\alpha T^a t^b\right) \tag{2.3}$$

where a and b are exponents that describe the volume effect. The diameter of the lens D and the sag height H were chosen as factors in the volume effect, described by

$$a(D, h) = 1.55 - 3.47\frac{h}{D} + 0.06\frac{D}{h} \tag{2.4}$$

$$b(D, h) = 0.27 - 1.36\frac{h}{D} + 0.04\frac{D}{h} \tag{2.5}$$

where -1.234×10^{-4} is used as a dimensional factor that changes with other conditions, such as photoresist material, substrate material, spin conditions, and humidity during lithography, baking, exposure, and development. The coefficient of determination (R2) of the estimated regression equation is 0.91, which shows that this function is a good fit for the data obtained. The volume change ratio for various photoresist volumes can be found using Eq. (2.4). Thus, the required initial volume of photoresist for a desired reflow lens can be determined, and the reflow process designed accordingly.

2.2.3.3 Verification of the Model

To verify the volume change ratio equation, a new microlens was designed and fabricated using Eq. (2.4). An aspheric lens was designed via the following equation:

$$z(r) = H + \frac{r^2/R_c}{1 + \sqrt{1 - (1 + K)r^2/R_c^2}} + Ar^4 \tag{2.6}$$

where R_c is the radius of curvature at the center of the lens, H is the sag height, k is the conic constant, and A is the aspheric coefficient.

Using Eq. (2.6), an aspherical microlens with a diameter of 98.6 µm, a focal length of 136.60 µm (refractive index 1.571), an effective pupil diameter of 85 µm, and a focal number of 1.61 was designed as follows:

$$z(r) = 0.018 - \frac{r^2/0.085}{1 + \sqrt{(1 + 0.3952)r^2/0.085^2}} - 246.325r^4 \tag{2.7}$$

The volume of this lens can be calculated from

$$V_{\text{lens}} = 2\pi \int_0^{D/2} z(r)r\,dr = 73886.66\ \mu\text{m}^3 \tag{2.8}$$

Setting H equal to 18 μm and D equal to 98.6 μm, the volume coefficients a and b were determined to be 1.26 and 0.25, respectively. A heating temperature of 150°C and a reflow time of 60 s were chosen as the reflow conditions. Utilizing Eq. (2.3), the volume change ratio was calculated as 0.8273.

From Eq. (2.2), the volume of the initial photoresist island may be calculated from

$$V_{tc} = \frac{V_{lens}}{f(T, t)} = 89310.60 \ \mu m^3 \tag{2.9}$$

Because the diameter of the reflow lens is not changed during the reflow process, the diameter of the initial photoresist island was set at 98.6 μm. At the design stage, the taper was chosen to be 15° since the taper of the photoresist island was close to 15° in most cases. Substituting these values into Eq. (2.1), the thickness of the initial photoresist island was calculated as 12.53 μm.

Experiments were performed using the designed values ($D_b = 98.6$ μm, $h = 12.53$ μm), and the volumes of the fabricated photoresist islands were measured. The average volume of the photoresist islands was 89300 μm³. Tests of the reflow process were performed on the fabricated photoresist islands. The heating temperature and reflow time were set at 150°C and 60 s, respectively, as determined at the design stage. After reflow, the profiles of the reflow microlenses were measured. Table 2.1 compares the profiles, volumes, and optical properties of the designed lens and the reflow lenses. The focal lengths and RMS wavefront aberrations of the reflow lenses were obtained by Code V (Optical Research Associates, US) using the measured profiles and a refractive index of 1.571. The sag heights and radii of curvature of the reflow lenses deviated from those of the designed lens by less than 0.5 and 3 μm, respectively. Therefore, the volumes and focal lengths of the reflow lenses deviated from those of the designed lens by less than 1.5% and 2.5%, respectively. Figure 2.18 compares the profiles of the designed and fabricated

TABLE 2.1 Comparisons of Various Parameters between Design Lens and Reflow Lens

	Sag Height	Radius of Curvature	Volume(error)	Focal Length(error)
Design	18	85	73886.66	136.60
Test 1	18.14	85.3	73280.34 (0.8%)	137.08 (−0.3%)
Test 2	17.69	87.2	72924.54 (1.3%)	140.13 (−2.5%)
Test 3	17.90	87.1	74854.07 (−1.3%)	139.97 (−2.5%)

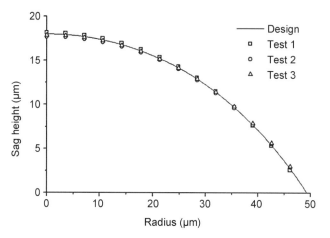

FIGURE 2.18 Comparison of lens profiles for a designed lens and fabricated lenses.

lenses. The deviation from the design profile was calculated by subtracting the profile of the fabricated microlens from the design profile of the aspherical lens, as shown in Figure 2.19. The deviation from the designed profile was less than 0.5 μm. These errors are attributed to other process conditions, such as the level of moisture and solvent present, atmospheric temperature and humidity, and the photolithographic conditions. However, these errors are small enough to be ignored, and the results demonstrate the validity of the present volume change ratio equation. Using this equation, reflow lenses can be fabricated more precisely, and the development time required to obtain a reflow lens with a desired profile can be reduced.

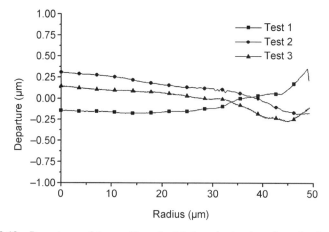

FIGURE 2.19 Departures of the profiles of a fabricated microlens from the design profile.

2.2.4 Laser Interference Lithography

2.2.4.1 *Theory of Laser Interference Lithography* Laser interference lithography, which uses optical interference, is capable of fabricating periodic and precise submicron gratings. Examples of its usage are found in the production of various filters and holographic optical elements (HOEs). Optical interference theory is used to calculate the periodicity of the gratings [63].

In free space, Maxwell's equations are given by

$$\nabla^2 E = \varepsilon_0 \mu_0 \frac{\partial^2 E}{\partial t^2}$$

$$\nabla^2 B = \varepsilon_0 \mu_0 \frac{\partial^2 B}{\partial t^2}$$

where E is the electric field, B is the magnetic field, ε_0 is the permittivity, and μ_0 is the permeability. For a plane wave, the electric field is expressed by

$$\vec{E} = \hat{e}|E|\exp\left[i\left(\vec{k} \cdot \vec{r} - \varpi t + \phi\right)\right] \tag{2.10}$$

The optical intensity (or irradiance) is defined as

$$I \equiv \left\langle c^2 \varepsilon_0 |\vec{E} \times \vec{B}| \right\rangle = c\varepsilon_0 \langle \vec{E} \cdot \vec{E} \rangle \tag{2.11}$$

From Eq. (2.10),

$$I = c\varepsilon_0 |E|^2 \left\langle \cos^2\left(\vec{k} \cdot \vec{r} - \varpi t\right) \right\rangle = \frac{c\varepsilon_0}{2} |E|^2 \tag{2.12}$$

If two plane waves are polarized in the same direction, they can be written as

$$\vec{E}_1 = \hat{e}_0|E_1|\exp\left[i\left(\vec{k}_1 \cdot \vec{r} - \varpi_1 t + \phi_1\right)\right]$$

$$\vec{E}_2 = \hat{e}_0|E_2|\exp\left[i\left(\vec{k}_2 \cdot \vec{r} - \varpi_2 t + \phi_2\right)\right] \tag{2.13}$$

The electric field at point P on a plane is

$$\vec{E} = \vec{E}_1 + \vec{E}_2 \tag{2.14}$$

Combining Eqs. (2.12)–(2.14), the intensity is then given by

$$\begin{aligned}
I &= c\varepsilon_0 \left\langle \left(\vec{E}_1 + \vec{E}_2\right) \cdot \left(\vec{E}_1 + \vec{E}_2\right) \right\rangle \\
&= c\varepsilon_0 \left\langle \left(\vec{E}_1 \cdot \vec{E}_1 + \vec{E}_2 \cdot \vec{E}_2 + 2\vec{E}_1 \cdot \vec{E}_2\right) \right\rangle \\
&= c\varepsilon_0 \left\langle \left(\vec{E}_1 \cdot \vec{E}_1 + \vec{E}_2 \cdot \vec{E}_2 + 2\vec{E}_1 \cdot \vec{E}_2\right) \right\rangle c\varepsilon_0 \left(\langle \vec{E}_1 \cdot \vec{E}_1 \rangle + \langle \vec{E}_2 \cdot \vec{E}_2 \rangle \right)
\end{aligned}$$

$$+2\langle \vec{E}_1 \cdot \vec{E}_2 \rangle)$$

$$= c\varepsilon_0 \left[\frac{1}{2}|E_1|^2 + \frac{1}{2}|E_2|^2 + |E_1||E_2|\cos\left\{ \left(\vec{k}_1 - \vec{k}_2\right) \cdot \vec{r} + (\phi_1 - \phi_2) \right\} \right]$$

$$= I_1 + I_2 + 2\sqrt{I_1 I_2} \cos\left\{ \left(\vec{k}_1 - \vec{k}_2\right) \cdot \vec{r} + (\phi_1 - \phi_2) \right\} \qquad (2.15)$$

When two plane waves strike a photoresist on a substrate at the same angle, the wave propagation vectors are

$$\vec{k}_1 = k_0\left(\hat{i}\sin\theta - \hat{j}\cos\theta\right)$$
$$\vec{k}_2 = k_0\left(-\hat{i}\sin\theta - \hat{j}\cos\theta\right) \qquad (2.16)$$

where θ is the incident angle and k_0 is the propagation number ($k_0 = 2\pi/\lambda$). We can take the position vector at point P, where $\vec{r} = \hat{i}x + \hat{j}y$, and transform it into Eq. (2.16):

$$\left(\vec{k}_1 - \vec{k}_2\right) \cdot \vec{r} = \left(\hat{i}2k_0\sin\theta\right) \cdot \left(\hat{i}x + \hat{j}y\right) = 2k_0 x \sin\theta \qquad (2.17)$$

Hence, Eqs. (2.16) and (2.17) become simply

$$I = I_1 + I_2 + 2\sqrt{I_1 I_2} \cos(2k_0 x \sin\theta + \phi_1 - \phi_2) \qquad (2.18)$$

From Eq. (2.18), the periodicity p can be determined. Figure 2.20 shows the normalized intensity of one-dimensional gratings and two-dimensional crossed gratings with determining constants k_0 and θ. Therefore,

$$2k_0 p \sin\theta = 2\pi$$
$$2\pi = \frac{\pi}{k_0 \sin\theta} = \frac{\lambda}{2\sin\theta} \qquad (2.19)$$

The periodicity p of the 1D gratings or the 2D crossed gratings can be varied by changing the angle θ between the two respective laser beams and the substrate normal via $p = \lambda/(2\sin\theta)$.

2.2.4.2 Simulation of Laser Interference Lithography
Figures 2.21 and 2.22 show the simulation results for photoresist patterning by laser interference lithography. Figure 2.21 shows the simulated intensity profile distributions in the photoresist layer. In the exposure stage, the wavelength of the laser beam, the beam intensity, and the refractive index of the photoresist were set to 441.6 nm, 0.31 mW/cm², and 1.54, respectively.

(a)

Normalized
exposure
energy

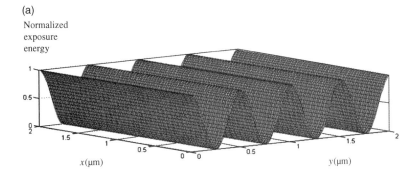

$x(\mu m)$ $y(\mu m)$

(b)

Normalized
exposure
energy

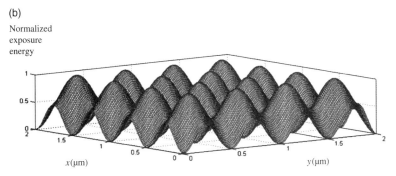

$x(\mu m)$ $y(\mu m)$

FIGURE 2.20 3D images of optical intensity of (a) 1D gratings with 1 μm pitch and (b) 2D crossed gratings with 1 μm pitch.

The incident angle was set to (a) 13° and (b) 3.17°. In both cases, the absorption coefficient and reflectance were also considered. Figure 2.22 shows the simulated photoresist surface profiles in the development stage as a function of development time at 15 s intervals. In this simulation, the dissolution rate was assumed to be nonlinear with respect to the exposure energy. The simulation results were used as guidelines to fabricate the master nanopatterns.

2.2.4.3 Experimental Setup

The periodic nano/micrograph were generated by optical interference. For this purpose, two coherent laser beams were superimposed to form sinusoidal interference patterns. A 4 inch silicon wafer coated with photoresist was exposed to these interference patterns. Figure 2.23 shows the laser interference lithography system used in the experiment. The light source for exposure was a He–Cd laser with a wavelength of 441.6 nm, manufactured by Kimmon Electric Co. Ltd., in Japan. To remove the effects of vibration and air particles, we installed the optical components on an antivibration table in a clean booth. The beam

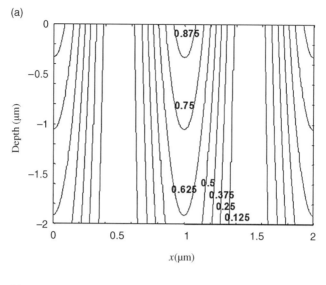

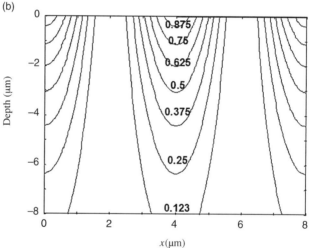

FIGURE 2.21   Simulated intensity profile distribution in photoresist. Exposure condition for interference pattern of (a) $13°$ incidental angle, $1\,\mu m$ pitch; (b) $3.17°$ incidental angle, $4\,\mu m$ pitch.

emitted from the laser was passed through the shutter and divided into two laser beams by a beam splitter. Both beams were then reflected by the internal reflection mirrors, and the intensity of each laser beam was modulated using a neutral density filter. Next, the laser beams were expanded via a spatial filter. The resulting standing wave grating image, with period $p = \lambda/(2\sin\theta)$, was captured in a photosensitive resist.

(a)

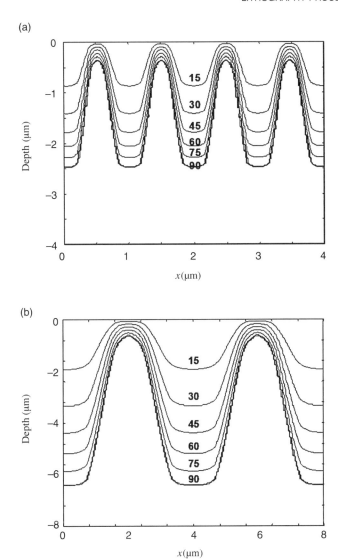

(b)

FIGURE 2.22 Simulated photoresist surface profile as a function of development time with 10 s interval. Exposure condition for interference pattern of (a) 13° incidental angle, 1 μm pitch and (b) 3.17° incidental angle, 4 μm pitch.

2.2.4.4 Fabrication of Nanostructures Using Laser Interference Lithography under Different Process Conditions The fabrication

process for the periodic micro/nanostructures is now described. Nanofabrication of line patterns and dot patterns on a substrate by laser interference lithography is accomplished in four steps: (1) wafer cleaning and

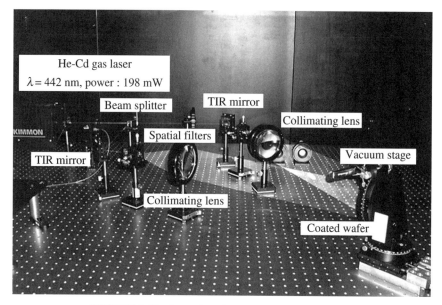

FIGURE 2.23   Laser interference lithography system.

HMDS deposition, (2) photoresist coating, (3) laser interference exposure to produce the patterns, and (4) developing. Before depositing the photoresist, the samples were cleaned by a conventional method using acetone, methyl alcohol, and deionized (DI) water. Then, 1.2 µm of positive photoresist was coated on the silicon substrate by a spin coater at 4000 rpm. In this study, AZ6612K was used as an i-line photoresist. The 4 inch silicon wafer was coated with photoresist and exposed to the laser interference to form optical patterns. (The two halves of the laser beam interfere and form grating patterns with periodicity p, which depends on the incident angle of the laser beam to the photoresist.) To fabricate dot patterns with a 4 µm pitch, the incident angle was set to $3.17°$. The dot patterns were fabricated by exposing the line and space patterns after a $90°$ rotation. It was necessary to optimize the laser interference lithography process to obtain the maximum dot height and a clear circular pattern since the lithographed pattern was used as the mask in the RIE process described in Section 2.1. In the laser interference lithography process, the exposure intensity and exposure time are considered to be the dominant processing parameters that affect the pattern shape. An exposure time of 140 s (suggested by the simulation) resulted in the maximum dot pattern height. However, these patterns were limited to usage as membrane filters since they were interconnected, as shown in Figure 2.24a. Thus, the exposure time was experimentally optimized to 180 s to obtain clear dot patterns suitable for filter applications, as indicated in Figure 2.24b. The intensity of each laser beam was $0.31\,\mathrm{mW/cm^2}$, and the developing time was 50 s. Figure 2.24b shows a scanning electron microscope image of the nanodot array with a height of 500 nm. The dot size was very uniform over the entire $10\,\mathrm{mm} \times 10\,\mathrm{mm}$ area.

(a)

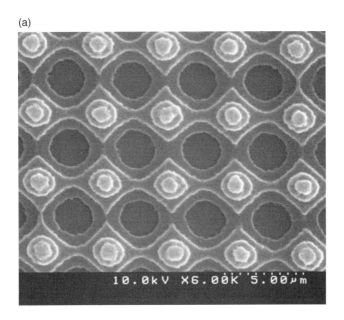

(b)

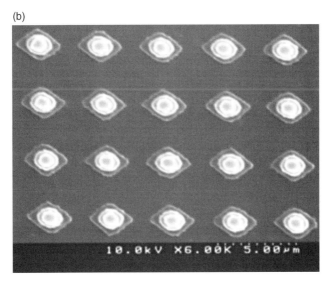

FIGURE 2.24 SEM images of (a) PR pattern with 140 s exposure time and (b) PR pattern with 180 s exposure time with 4 μm pitch.

2.3   ELECTROFORMING PROCESSES

A silicon or glass mold insert is too brittle to be used in compression molding or injection molding for mass production, where high-pressure shocks are repeatedly applied to the mold cavity. A metallic mold insert is a possible solution to this problem, and in this case, either mechanical machining or electroforming can be used to create the micro/nanopatterns [64]. It is more reasonable to fabricate the mold insert by electroforming if the required patterns are to be small and dense. To fabricate a mold insert via the electroforming process, a mother of the same shape as the final molded component is first required. Any of the established fabrication methods for micro/nanocomponents can be used to fabricate the mother. Nickel electroforming is a well-established technique for fabricating a mold insert for the optical disk injection molding process [65,66]. Initial deposition of a seed layer is required for electroforming since the surface of the mother is generally a nonconducting material. Evaporation, sputtering, and nonelectrodeposition can be used for the metal deposition. Nickel, gold, or silver is generally used as a seed layer material since these metals possess desirable surface properties such as high hardness and thermal stability. After the seed layer is deposited, the nickel is electroformed. Figure 2.25 shows a schematic diagram of an electroforming system. The residual stresses in the electroformed layer can be suppressed by controlling the evaporation and electroforming processes, and by the use of properly designed jig. After electroforming, the back of the electroformed mold insert is polished to obtain the desired thickness and flatness. Then, the mother is separated from the metal mold insert. In the next section, we review the theory of electroforming and various experimental results.

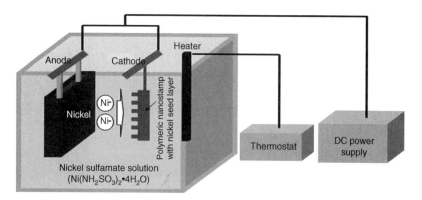

FIGURE 2.25   A schematic diagram of the electroforming system.

2.3.1 Theory of Electroforming Process

Electroforming is the process of synthesizing a metal object by controlling the electrodeposition of metal via an electrolytic solution. A metal layer is built up on a metal surface or on any surface that has been rendered electroconductive through the application of a paint containing metal particles [67]. The master pattern is placed in electrical contact with the cathode, and the target material for fabrication is placed in contact with the anode. The material at the anode is oxidized and dissolved. The cations are reduced at the cathode, and metal is deposited on the master. The thickness of the electroformed nickel layer can be controlled by the current density and the deposition time [68]. In this study, nickel was selected for the electroforming layer material because of its hardness and durability. Nickel sulfamate solution $(Ni(NH_2SO_3)_2 \cdot 4H_2O)$ was used to fabricate the metallic mold.

2.3.2 Electroforming Results

2.3.2.1 Metallic Mold for a Microlens Array
An E-beam evaporation system was used to deposit the seed layer on the mother lens, which was fabricated via the reflow process. Nickel was chosen as the seed layer material because it possesses desirable surface properties such as high hardness and thermal stability. The nickel was electroformed following the deposition of a seed layer with a thickness of $1500\,\text{Å}$. Commercial nickel sulfamate solution was used as an electrolyte. The electroforming bath consisted of a 66 l PVC bath, a power supply, an air pump for agitation, a filter, and a thermostat for temperature control. The bath was maintained at 43–45°C during electroforming by a current density of 10–20 mA/cm^2, yielding an approximate deposition rate of 1–2 μm/min. The pH was maintained at 3.8–4.4. We controlled the evaporation and electroforming processes, and used a properly designed jig to suppress the development of residual stresses in the electroformed layer. After the electroforming was completed, the backside of the electroformed mold was polished to obtain the desired mold insert thickness and flatness, and the silicon wafer and photoresist were then removed. Figure 2.26 shows SEM images of the electroformed nickel mold. The diameters of the concave surfaces were 36 and 96 μm, respectively, and the pitch of the array was 250 μm in both cases.

2.3.2.2 Metallic Mold for Patterned Media
A metallic mold with full tracks of nanohole patterns was fabricated using a seed layer deposition process and a subsequent electroforming process [69]. Sputtering was used to deposit the conductive seed layer on the polymeric nanomaster. It should be noted that

(a)

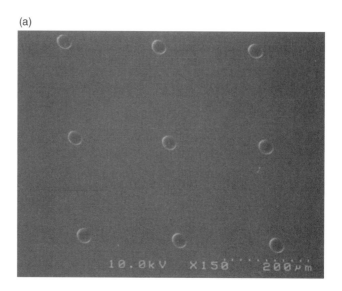

(b)

FIGURE 2.26 SEM images of metallic stamp of microlens: (a) cavity diameter of 36 μm; (b) cavity diameter of 96 μm. Reprinted with permission from Ref. [64]. Copyright 2003, Institute of physics science.

the seed layer determines the surface quality and mechanical properties of the metallic mold. Nickel was used as the seed layer material because of its hardness, durability, and adhesion properties. During the sputtering process, deposited metallic particles can attack the nanopillar patterns on a polymeric master, which may result in deterioration of the pattern fidelity of a metallic nanomold. For this reason, the deposition conditions were precisely optimized.

TABLE 2.2 Optimized Processing Parameters for Sputtering of the Nickel Seed Layer on Polymer Master

Parameter	Value
Base pressure (Torr)	1.3×10^{-7}
Working pressure (Torr)	3.0×10^{-3}
Deposition rate of nickel (Å/s)	1.38
Temperature (°C)	20.0
Direct current (kW)	0.1
Applied voltage (V)	200
Flow rate of argon gas (cm^3/min)	22.2

Table 2.2 lists the controlled processing parameters of the sputtering process for the deposition of a nickel seed layer on a polymeric master.

After the seed layer was deposited on the master, electroforming was carried out using nickel sulfamate solution to fabricate the metallic mold with full tracks of nanohole patterns. The electroforming process was controlled to suppress the separation of the seed layer and the development of residual stresses in the electroformed layer. The temperature of the electroforming bath was maintained at 53°C, and the pH was maintained at 3.84. During the ramping period of 450 s, the current density was increased from 0 to 47.2 mA/cm^2 for stable formation of the nickel-electroformed layer on the nickel seed layer. After the construction of the initial electroformed layer during the ramping period, the current density was maintained at a constant value of 47.2 mA/cm^2. The deposition rate was precisely controlled at 12.2 nm/s to reduce the residual stresses and shrinkage in the electroformed layer. The processing parameters for nickel electroforming are summarized in Table 2.3. The resulting thickness of the nickel mold was 330 μm. The thickness of the nickel mold was reduced to 300 μm after back polishing, and the diameter was 80 mm after wire cutting. The hardness of the metallic mold was analyzed using a standard

TABLE 2.3 Optimized Processing Parameters for Nickel Nanoelectroforming

Parameter	Value
Temperature (°C)	53
pH	3.84
Current density (mA/cm^2)	47.2
Ramping time (s)	450
Concentration of $Ni(NH_2SO_3)_2 \cdot 4H_2O$ (g/L)	350 ± 5
Concentration of $NiCl_2 \cdot 6\,H_2O$ (g/L)	5
Concentration of H_2BO_3(g/L)	35

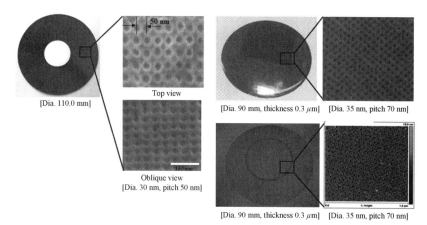

FIGURE 2.27 Metallic mold of patterned media. Reprinted with permission from Ref. [52]. Copyright 2009, Institute of Electrical and Electronics Engineers.

Rockwell-C hardness tester (ATK-F1000). The hardness of the metallic mold was measured at 40.7 HRC.

Figure 2.27 shows the fabrication results for the metallic mold with full tracks (track width of 1.3 mm and track inner diameter of 40 mm). Figure 2.27 also shows the details of the nanohole patterns (diameter of 30 nm, pitch of 50 nm, and depth of 50 nm). Full tracks of the nanohole patterns were successfully transferred onto the metallic mold from the full tracks of the nanopillar patterns on the polymeric master.

REFERENCES

1. J. R. Sheats and B. W. Smith (1998) *Microlithography Science and Technology*, Marcel Dekker Inc.

2. N. Taniguchi (1996) *Nanotechnology Integrated Processing Systems for Ultra-Precision and Ultra-Fine Products*, Oxford University Press.

3. J. A. Menapace, S. N. Dixit, F. Y. Génin, and W. F. Brocious (2004) Magnetorheological finishing for imprinting continuous phase plate structure onto optical surfaces. *Proceedings of SPIE*, 5273, 220.

4. M. Cho, H. Lim, C. Lee, B. Cho, and J. Park (2008) Fabrication of stainless steel mold using electrochemical fabrication method for microfluidic biochip. *Japanese Journal of Applied Physics*, 47, 6, 5217–5220.

5. P. R. Krauss and S. Y. Chou (1997) Nano-compact disks with 400 Gbit/in^2 storage density fabricated using nanoimprint lithography and read with proximal probe. *Applied Physics Letters*, 71, 21.

6. Z. D. Popovic, R. A. Sprague, and G. A. Neville Connell (1988) Technique for monolithic fabrication of microlens arrays. *Applied Optics*, 27, 1281.

7. M. C. Hutley, R. F. Stevens, and D. Daly (1990) The manufacture of microlens arrays and fan-out gratings in photoresist. *IEE Colloquium on Optical Connection, Optical Connection and Switching Network for Communication and Computing*, The Institution of Electrical Engineers, London, UK, 11/ 1.

8. D. Daly, R. F. Stevens, M. C. Hutley, and N. Davies (1990) The manufacture of microlenses by melting photoresist. *Measurement Science and Technology*, 1, 759.

9. N. Nordman and O. Nordman (2001) Optical properties of two photoresists designed for charge-coupled device microlenses. *Optical Engineering*, 40, 2572.

10. Y. S. Lin, C. T. Pan, K. L. Lin, S. C. Chen, J. J. Yang, and J. R. Yang (2001) Polyimide as the pedestal of batch fabricated micro-ball lens and micro-mushroom array category: micro opto electro mechanical systems. *14th IEEE Int. Conf. Micro Electro Mechanical Systems*, IEEE Instrumentation and Measurement Society, Interlaken, Switzerland, 337.

11. P. Heremans, J. Genoe, M. Kuijk, R. Vounckx, and G. Borghs (1997) Mushroom microlenses: optimized microlenses by reflow of multiple layers of photoresist. *IEEE Photonics Technology*, 9, 1041.

12. T. Fujita, H. Nishihara, and J. Koyama (1981) Fabrication of micro lenses using electron-beam lithography. *Optical Letters*, 6, 613.

13. T. Fujita, H. Nishihara, and J. Koyama (1982) Blazed gratings and Fresnel lenses fabricated by electron-beam lithography. *Optical Letters*, 7, 578.

14. S. V. Babin and V. A. Danilov (1998) Data preparation and fabrication of DOE using electron-beam lithography. *Optics and Lasers in Engineering*, 29, 307.

15. S. Mihailov and S. Lazare (1993) Fabrication of refractive microlens arrays by excimer laser ablation of amorphous Teflon. *Applied Optics*, 32, 6211.

16. H. M. Presby, A. F. Benner, and C. A. Edwards (1990) Laser micromachining of efficient fiber microlenses. *Applied Optics*, 29, 2692.

17. I. Zergioti, S. Mailis, N. Vainos, A. Ikiades, and C. P. Grigoropoulos (1999) Microprinting and microetching of diffractive structures using ultrashort laser pulses. *Applied Surface Science*, 82, 138–139.

18. K. Zimmer, D. Hirsch, and F. Bigl (1996) Excimer laser machining for the fabrication of analogous microstructures. *Applied Surface Science*, 425, 96–98.

19. N. George and J. W. Matthews (1966) Holographic diffraction gratings. *Applied Physics Letters*, 9, 212.

20. S. T. Chou, P. R. Krauss, and P. J. Renstrom (1995) Imprint of sub-25 nm via and trenches in polymers. *Applied Physics Letters*, 67, 3114.

21. F. W. Ostermayer, Jr., P. A. Kohl, and R. H. Burton (1983) Photoelectrochemical etching of integral lenses on InGaAsP/InP light-emitting diodes. *Applied Physics Letters*, 43, 642.

22. D. A. Fletcher, K. B. Crozier, K. W. Guarini, S. C. Minne, G. S. Kino, C. F. Quate, and K. E. Goodson (2001) Microfabricated silicon solid immersion lens. *Journal of Microelectromechanical Systems*, 10, 450.

23. L. Erdmann and D. Efferenn (1997) Technique for monolithic fabrication of silicon microlenses with selectable rim angles. *Optical Engineering*, 36, 1094.

24. M. B. Stern, M. Holz, S. S. Medeiros, and R. E. Knowlden (1991) Fabricating binary optics: process variables critical to optical efficiency. *Journal of Vacuum Science & Technology B*, 9, 3117.

25. J. Vukusic, J. Bengtsson, M. Ghisoni, A. Larsson, C. F. Carlstom, and G. Landgren (2000) Fabrication and characterization of diffractive optical elements in InP for monolithic integration with surface-emitting components. *Applied Optics*, 39, 398.

26. J. Bengtsson, N. Eriksson, and A. Larsson (1996) Small-feature-size fan-out kinoform etched in GaAs. *Applied Optics*, 35, 801.

27. J. N. Mait, A. Scherer, O. Dial, D. W. Prather, and X. Gao (2000) Diffractive lens fabricated with binary features less than 60 nm. *Optical Letters*, 25, 381.

28. R. Steingruber and M. Ferstl (2000) Three-dimensioal microstrusture elements fabricated by electron beam lithography and dry etching technique. *Microelectronic Engineering*, 53, 539.

29. S. Moon, S. Ahn, S. Kang, D. Choi, and T. Je (2001) Fabrication of refractive and diffractive plastic micro-optical components using micro-compression molding. *Proceedings of SPIE*, 4592, 140.

30. L. Lee, M. Majadou, K. Koelling, S. Daunert, S. Lai, C. Koh, Y. Juang, Y. Lu, and L. Yu (2001) Design and fabrication of CD-like microfluidic platforms for diagnostics; polymer-based microfabrication. *Biomed, Biomedical Microdevices* 3(4), 339–351.

31. C. G. Blough, M. Rossi, S. K. Mack, and R. L. Michaels (1997) Single-point diamond turning and replication of visible and near-infrared diffractive optical elements. *Applied Optics*, 36, 20.

32. Y. Yamagata and S. Morita (2001) Fabrication of blazed holographic optical element by ultrahigh-precision cutting. *RIKEN Review*, 34.

33. M. B. Fleming, and M. C. Hutley (1997) Blazed diffractive optics. *Applied Optics*, 36, 4635.

34. S. Kang (2004) Replication technology for micro/nano optical components, *Japanese Journal of Applied Physics*, 43(8b), 5706–5716.

35. H. M. Presby, A. F. Benner, and C. A. Edwards (1990) Laser micromachining of efficient fiber microlenses. *Applied Optics*, 29, 2692.

36. I. Zergioti, S. Mailis, N. A. Vainos, A. Ikiades, C. P. Grigoropoulos, and C. Fotakis (1999) Microprinting and microetching of diffractive structures using ultrashort laser pulses. *Applied Surface Science*, 82, 138–139.

37. K. Zimmer, D. Hirsch, and F. Bigl (1996) Excimer laser machining for the fabrication of analogous microstructures. *Applied Surface Science*, 425, 96–98.

38. Stephen Mihailov and Sylvain Lazare (1993). Fabrication of refractive microlens arrays by excimer laser ablation of amorphous Teflon. *Applied Optics*, 32, 6211.

39. A. Braun, K. Zimmer, B. Hoˆsselbarth, J. Meinhardt, F. Bigl and R. Mehnert (1998). Excimer laser micromachining and replication of 3D optical surfaces. *Applied Surface Science*, 911, 127–129.

40. M. B. Stern and T. Jay (1994) Dry etching for coherent refractive microlens arrays. *Optical Engineering*, 33, 3547.

41. S. Moon, S. Kang, and J. Bu (2002) Fabrication of polymeric microlens of hemispherical shape using micromolding. *Optical Engineering*, 41, 2267.

42. S. Kim, D. Kim, and S. Kang (2003) Replication of micro-optical components by ultraviolet-molding process. *Journal of Microlithography, Microfabrication, and Microsystems*, 2, 356.

43. N. Dumbravescu and M. Ilie (1999) Replication of diffractive gratings using embossing into UV-cured photo-polymers. *Micromachine Technology for Diffractive and Holographic Optics, SPIE*, 3879, 206–213.

44. J. Taniguchi, Y. Tokono, I. Miyamoto, M. Komuro, and H. Hiroshima (2002) Diamond nanoimprint lithography. *Nanotechnology*, 13, 592.

45. D.R. Harriott (1982) Electron beam lithography. *Journal of Vacuum Science & Technology*, 20, 3.

46. C. Vieu, F. Carcenac, A. Pépin, Y. Chen, M. Mejias, A. Lebib, L. Manin-Ferlazzo, L. Couraud, and H. Launois (2000) Electron beam lithography: resolution limits and applications. *Applied Surface Science*, 164, 111–117.

47. A. N. Broers, A.C.F. Hoole, and J. M. Ryan (1996) Electron beam lithography - Resolution limits. *Microelectronic Engineering*, 32, 131–142.

48. K. Masahiro, S. Megumi, H. Kazunobu, H. Yasuo, K. Osamu, K. Hiroaki, K. Masaki, I. Tetsuya, and K. Kazumi (2005) Electron beam recording beyond 200 Gbit/in^2 density for next generation optical disk mastering. *Japanese Journal of Applied Physics*, 44 (5B) 3578–3582.

49. T. M. Smeeton, M. J. Kappers, J. S. Barnard, M. E. Vickers, and C. J. Humphreys (2003) Electron-beam-induced strain within InGaN quantum wells: false indium "cluster" detection in the transmission electron microscope. *Applied Physics Letters*, 83, 26,

50. G. A. Gibson, A. Chaiken, K. Nauka, C. C. Yang, R. Davidson, and A. Holden (2005). Phase-change recording medium that enables ultrahigh-density electron-beam data storage. *Applied physics letters*, 86, 051902.

51. J. Han, J. Yang, S. Shin, Y. Kim, and S. Kang (2009). Design and fabrication of perpendicular patterned magnetic media using nanoimprinted nano pillar patterns. *IEEE Transactions on Magnetics*, 45, 5.

52. H. Kim, S. Shin, J. Han, J. Han, and S. Kang (2009) Fabrication of metallic nano stamp to replicate patterned substrate using electron-beam recording, nanoimprinting, and electroforming. *IEEE Transactions on Magnetics*, 45(5), 2288–2291.

53. J. Lim, M. Popall and S. Kang (2009) Application of organic/inorganic hybrid photopolymer to fabricate ultra-high-density patterned substrate. *IEEE Transactions on Magnetics*, 45(5), 2300–2303.

54. H. Miyajima and M. Mehregany (1995) High-aspect-ratio photolithography for MEMS applications. *Journal of Microelectromechanical Systems*, 4, 4.

55. P. F. Tian, P. E. Burrows, and S. R. Forrest (1997) Photolithographic patterning of vacuum-deposited organic light emitting devices. *Applied Physics Letters*, 71, 22.

56. J. A. Rogers, K. E. Paul, R. J. Jackman, and G. M. Whitesides (1997) Using an elastomeric phase mask for sub-100 nm photolithography in the optical near field. *Applied Physics Letters*, 70, 20.

57. M. B. Stern and T. R. Jay (1994) Dry etching for coherent refractive microlens array. *Optical Engineering*, 33, 3547–3551.

58. S. Moon, N. Lee, and S. Kang (2003) Fabrication of a microlens array using a micro-compression molding with an electroformed mold insert. *Journal of Micromechanics and Microengineering*, 13.

59. S. Sinzinger and J. Jahns (1999) *Microoptics*, Wiley, New York.

60. J. M. Shaw, M. A. Frisch, and F. H. Dill (1977) Thermal analysis of positive photoresist film by mass spectrometry. *IBM Journal of Resist Development*, 219–226.

61. T. R. Jay, M. B. Stern, and R. F. Knowlden (1992) Effect of refractive microlens array fabrication parameters on optical quality. *Miniature and Micro-Optics Proceedings of SPIE*, 1751, 179–183.

62. D. Daly (2001) *Microlens Arrays*, Taylor & Francis, New York.

63. P. K. Rastogi (1994) *Holographic Interferometry*, Springer, Berlin.

64. S. Moon, N. Lee, and S. Kang (2003) Fabrication of a microlens array using micro-compression molding with an electroformed mold insert. *Journal of Micromechanics and Microengineering*, 13, 98–103.

65. M. T. Gale (1997) Replication techniques for diffractive optical elements. *Microelectronic Engineering*, 34, 321–339.

66. K. Reimer, R. Engelke, U. Hofmann, P. Merz, K. T. Kohlmann, V. Platen, and B. Wagner (1999) Progress in gray-tone lithography and replication techniques for different materials. *Proceedings of SPIE*, 3879, 98.

67. K. Ansari, J. Kan, A. Bettiol and F. Watt, J (2006). Stamps for nanoimprint lithography fabricated by proton beam writing and nickel electroplating. *Journal of Micromechanics and Microengineering*, 16, 1967–1974.

68. H. Yang a, S. Kang (2000). Improvement of thickness uniformity in nickel electroforming for the LIGA process. *International Journal of Machine Tools & Manufacture*, 40, 1065–1072

69. S. Kim, J. Min, J. Lim, M. Choi, G. Han, and S. Kang (2008) Design and evaluation of wafer level UV-imprinted aspherical square microlens arrary to increase ray convergence efficiency for image sensor. *Japanese Journal of Applied Physics*, 47, 6667–6671.

CHAPTER 3

Modification of Mold Surface Properties

3.1 INTRODUCTION

Smaller pattern sizes in nanoimprinting, hot embossing, ultraviolet (UV) nanoimprinting, and injection molding for replicating ultrafine patterns lead to more difficulties in the interface between the mold and polymer due to the adhesion force between them. In thermal nano replication processes using metallic nanomold, as the features of the nanopatterns on the mold become smaller, one needs to raise the mold temperature. The reason is that the polymer above the glass transition temperature has higher fluidity than that below and, therefore, can fill the nanopatterns better. However, when the mold temperature increases excessively, the polymer may stick to the mold, deteriorating the replication quality as well as the surface quality of the replicated nanopatterns. Previous studies [1–4] have shown that the interfacial problems between the mold and the polymer deteriorate the surface profile and surface quality of replicated nanopatterns at high mold temperatures. As shown in Figure 3.1, when the replication temperature is 180°C, the polymer could not fill the submicron pattern in the mold completely; however, when the replication temperature is 240°C, the sticking problem between the mold and the polymer deteriorates the replication quality and surface quality of nanohole patterns in the replicated substrates. In the case of UV nanoimprinting of ultrafine patterns, in particular, the adhesion force of the curable polymer and the silicon mold cause defects in the replica due to sticking. These interfacial problems between the mold and the replicated patterns can be solved by the surface modifications. Among the surface modifications, coating an anti-adhesion layer to decrease the surface energy of the mold surface has been common practice to prevent sticking [5–8]. In nano replication applications, where the pattern sizes of the mold are

Micro/Nano Replication: Processes and Applications, First Edition. Shinill Kang.
© 2012 John Wiley & Sons, Inc. Published 2012 by John Wiley & Sons, Inc.

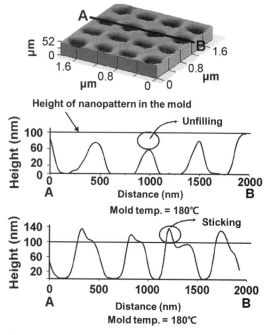

FIGURE 3.1 Replication qualities of replicated polymeric nanopatterns at different mold temperatures. Reprinted with permission from Ref. [9]. Copyright 2006, American Institute of Physics.

very small, conventional coating technology cannot be applied, because thickness of anti-adhesion layer on the mold may affect the shape of the nanopatterns in the mold.

Much research into anti-adhesion layers for replicating micro/nanopatterns has been conducted. The nano replication process requires decreasing the surface energy using methods such as diamond-like carbon, polytetrafluoroethylene, and self-assembled monolayers (SAM). SAM monomers are deposited selectively with the surface material of the mold and combine in an orderly manner on the surface. The combined monolayer is very thin, as little as a single molecule, and so has no influence on the shape of ultrafine patterns. In addition, the hydrophobic surface of the SAM decreases the surface energy of the mold surface and the adhesion force between the mold and polymer in the replication process. This makes the release process of the mold much easier and prevents damage and contamination of the mold and replica.

Many types of SAM are possible depending on the substrate. This study concentrates on silane- and thiol-based precursors. Silane-based precursor is deposited on silicon and quartz for nano replication and microelectromechanical system structures while thiol-based SAM are used with materials such as Au and Pt for biological applications. This chapter examines the two types of SAM for nano replication applications, such as injection molding using a silicon master for a nickel mold.

3.2 THIOL-BASED SELF-ASSEMBLED MONOLAYER

3.2.1 Thiol-Based Self-Assembled Monolayer and Deposition Process

SAM is one of the candidates to be used as an anti-adhesion layers in nano replication applications. SAM is physically and chemically stable and can modify the surface properties of the mold without affecting the shape of the nanopatterns on the mold, because the monomolecular of 2–3 nm thickness is adsorbed on the mold surface. Many researchers have investigated the SAM-deposition on the silicon or gold substrates [7,8,10–12]. SAM can be deposited on the silicon or gold substrates after simple pretreatments. However, silicon or gold substrates are not suitable materials for a mold in nano replication as mass production. Although nickel substrate is a superb material for a mold in nano replication applications, the oxide layer on the nickel surface deteriorated the anti-adhesion property of SAM-deposited nickel mold. However, it is not easy to carry out the pretreatment to reduce the oxide layer on the nickel mold. This is the reason why the studies about the deposition of SAM on the nickel substrate have focused on the pretreatment to reduce the oxide layer on the nickel surface.

Figure 3.2 shows the schematic structures of n-dodecanethiol ($CH_3(CH_2)_{11}SH$) SAM on the nickel mold. As shown in Figure 3.2,

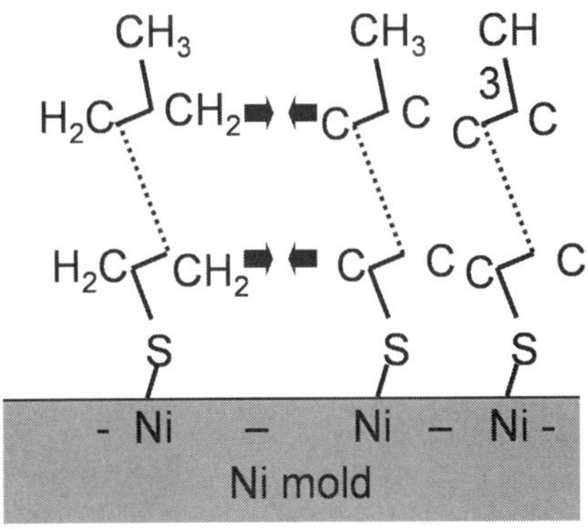

FIGURE 3.2 Schematic structure of n-dodecanethiol monolayer ($CH_3(CH_2)_{11}SH$) deposited on the nickel mold. Reprinted with permission from Ref. [9]. Copyright 2006, American Institute of Physics.

n-dodecanethiol SAM comprises 12 carbon chains and its reaction group, -SH, links with nickel atoms. Then, carbon chains of molecular, $-CH_2-$, keep close to other carbon chains by van der Waals force and these molecular combinations build the monolayer orderly on the nickel surface. And the function group, $-CH_3$, decreases the surface energy of the monolayer.

n-Dodecanethiol SAM is applied as an anti-adhesion layer to reduce the adhesion problem between the nickel mold and replicated nanopatterns in nano replication processes for the first time. Before depositing *n*-dodecanethiol SAM on the nickel mold, the nickel mold was pretreated to remove the oxide layer on the mold surface. Using the solution deposition method, *n*-dodecanethiol SAM as an anti-adhesion layer was deposited on the nickel mold. To examine the feasibility of the *n*-dodecanethiol SAM as an anti-adhesion layer, contact angle and the lateral friction force were measured at room temperature.

Before depositing SAM on the nickel mold, the nickel mold was pretreated by electrochemical reduction method, as shown in Figure 3.3a. After electrochemical pretreatment process, SAM was deposited on the nickel

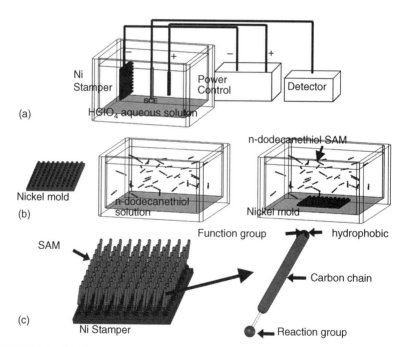

FIGURE 3.3   SAM deposition on the nickel mold using the solution deposition method. (a) Electrochemical pretreatment; (b) dipping nickel mold in the *n*-dodecanethiol solution; (c) organization of *n*-dodecanethiol monolayer. Reprinted with permission from Ref. [9]. Copyright 2006, American Institute of Physics.

mold using solution deposition method. *n*-Dodecanethiol monolayer was formed by immersion of the nickel substrates in the neat *n*-dodecanethiol solution, as shown in Figure 3.3b.

3.2.2 Experiment Results and Analysis

Verifying the SAM on the substrate is very difficult due to the extreme thinness of the monolayer. The SAM deposition can be studied using methods such as X-ray photoelectron spectroscopy (XPS) to analyze the elements, or contact angle and lateral force microscopy to examine the monolayer properties. In XPS, incident X-rays cause the photoemission of electrons from the sample surface, the energy of which is analyzed to obtain the binding energy of the emitted electrons. Binding energies are characteristic of each element and can be used for identification, while peak areas allow for quantification. Detection of the binding energy of the -SH element of *n*-dodecanethiol SAM indicates whether the SAM coating exists on the nickel substrate. Figure 3.4 shows an XPS analysis of nickel substrate with and without an SAM coating. Figure 3.4a shows no indication of sulfur in the binding energy region. In Figure 3.4b, the energy peak of the S_2^{2-} element (binding energy of 162.3 eV) indicates the presence of an SAM coating on the nickel substrate. This indicates that the *n*-dodecanethiol monolayer combines with nickel atoms of the substrate.

To verify the feasibility of the SAM-deposited nickel mold, contact angle was measured at room temperature, and the lateral friction force (LFF) was measured at different normal forces (4–12 nN). Table 3.1 summarizes the surface properties of a bare nickel mold and a nickel mold with a deposited SAM. As shown in Figure 3.5, the contact angle between nickel mold and deionized water increased from 70.37° to 109.22° after SAM deposition. Figure 3.6 shows the lateral friction forces for the bare nickel mold and the SAM-deposited nickel mold at different normal forces. In the case of SAM-deposited nickel mold, lateral friction forces at different normal forces decreased by 40%, compared to the bare nickel mold. As shown in Figure 3.6, as the normal force increased, SAM can reduce the lateral friction force markedly. This tendency of lateral friction forces at high normal forces indicates that SAM-deposited nickel mold reduced effectively the sticking force between nickel mold and replicated parts at the high replication pressure in the actual replication process. Additionally, these results showed that wet energy of the SAM-deposited mold markedly decreased from 24.26 to −23.97 mN/m due to SAM deposition. LFF and contact angle results indicated that the molecules of the monolayer were constructed uniformly on the nickel mold and the -CH$_3$ group of the *n*-dodecanethiol SAM reduces surface energy effectively. Therefore, these results imply the

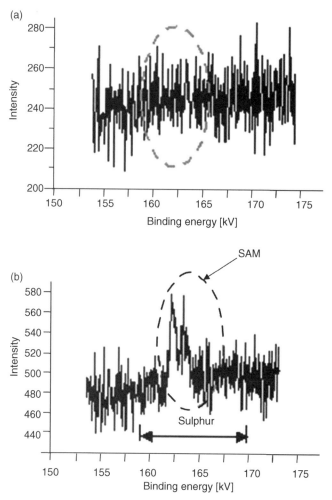

FIGURE 3.4 Binding energy data of XPS for analysis of monolayer deposited on nickel substrate (a) bare nickel substrate and (b) SAM-deposited nickel substrate.

feasibility of SAM as an anti-adhesion layer for the nano replication process using a nickel mold.

3.2.3 The Changing Properties of SAM at Actual Replication Environment

A nickel mold with a deposited SAM had a lower surface energy than a bare nickel mold, indicating the possibility of using an SAM as an anti-adhesion layer on a metallic mold for nano replication processes. However, these replication processes are actually performed under various process and environmental conditions, as summarized in Table 3.2. Figure 3.7 shows

TABLE 3.1 **Comparisons of Surface Properties between the Bare Nickel Mold and the SAM-Deposited Nickel Mold for 10 nN Normal Force of LFF**

Parameter	Bare Nickel Mold	SAM-Deposited Nickel Mold
Contact angle ($^\circ$)	70.37	109.22
Wetting energy (mN/m)	24.46	−23.97
Lateral friction force (eV)	0.0969	0.0605

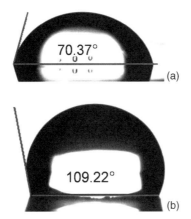

FIGURE 3.5 (a) Contact angle of the bare nickel mold, 70.37°, (b) contact angle of the SAM-deposited nickel mold, 109.22°. Reprinted with permission from Ref. [9]. Copyright 2006, American Institute of Physics.

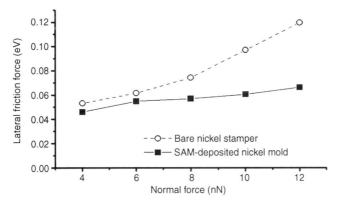

FIGURE 3.6 Lateral friction force for different normal forces for the bare nickel mold and the SAM-deposited nickel mold. Reprinted with permission from Ref. [9]. Copyright 2006, American Institute of Physics.

TABLE 3.2   Parameters of Actual Replication Processes

	Injection Molding	Hot Embossing	UV Imprinting
Temperature (°C)	150–300	100–210	Room temperature
Pressure (MPa)	8–10/DVD-RAM	8/10 nm line width	0.1
UV dose (mJ/cm^2)	—	—	400

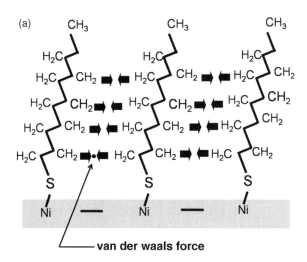

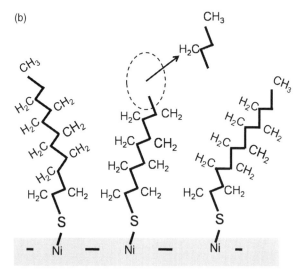

FIGURE 3.7   (a) The structure of self-assembled monolayer bonded by van der Waals force. (b) The structure of self-assembled monolayer destroyed by the heat energy or ultraviolet light.

that the SAM is damaged by heat energy and UV light. The SAM is composed of carbon chains held together by van der Waals force, as shown in Figure 3.7a. This force is so weak that the monolayers it holds together can be damaged by heat energy and strong light, as shown in Figure 3.7b [13,14]. Nano replication requires a high temperature to melt the polymer resin, while injection molding, hot embossing, and microcompression molding require high pressure to fill the patterns with resin. In addition, nano-UV imprinting requires UV light to cure the polymer. These parameters in various replication processes can affect the composition of the SAM.

To analyze the change in the surface energy of an SAM-deposited nickel mold under these conditions, the contact angle and lateral friction force were measured at the actual process temperature for nanocompression molding and injection molding, and at the actual UV dose for nano-UV imprinting.

The actual process temperatures selected were 100, 150, 200, 250, and 300°C. In this range, the SAM on the nickel substrate was heated for 5 min to the selected maximum replication temperature using a hotplate to control the experimental environment. The mold was then cooled to room temperature, and the contact angle and lateral fiction force were measured at that temperature. One SAM-deposited mold was tested at all the maximum temperatures to obtain consistent results for the various maximum replication temperatures because each SAM-deposited mold had a different contact angle. Otherwise, all the molds were tested at 100°C and then at increasing temperatures in 50°C intervals.

Figures 3.8 and 3.9 show the contact angle between deionized water and the SAM, and the lateral friction force of the SAM deposited on the nickel mold for different maximum replication temperatures. The contact angle remained constant up to a maximum replication temperature of 200°C, and then decreased significantly above 200°C as shown in Figure 3.9a. Figure 3.9b shows that the lateral friction force remained constant up to 200°C, similar to the contact angle, and increased significantly above 200°C. The monolayer is composed by carbon chains of alkanethiol bound by van der Waals force. As shown in Figure 3.9b, the lateral friction force increased slightly up to 200°C, indicating that these carbon chains were gradually damaged as the temperature increased. The vibration energy of atoms increases as the temperature increases. This increasing vibration breaks down the carbon bonding and causes defects in the monolayer or changes in the structure of the monomers to the point where the hydrophobic property of the SAM deteriorates.

UV exposure times of 450, 900, 1350, and 1800 s were selected to simulate the UV dose for actual UV imprinting. The SAM-deposited nickel mold was exposed to ultraviolet light for time increments of 450 s. The contact angle and lateral friction force were then measured at room temperature.

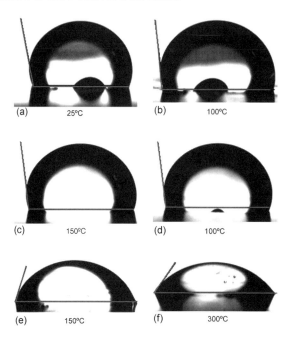

FIGURE 3.8 Images of water contact angles for different maximum replication temperatures; SAM-deposited nickel mold cooled at room temperature after heating at different maximum replication temperatures for 5 min.

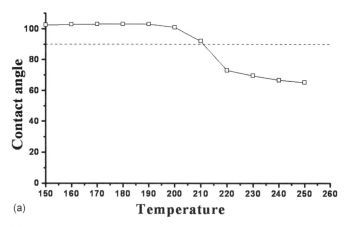

FIGURE 3.9 (a) Water contact angle for the range from 150 to 250°C, (b) water contact angle and lateral friction force as function of maximum replication temperature; SAM-deposited nickel mold was cooled at room temperature after heating at different maximum replication temperatures for 5 min.

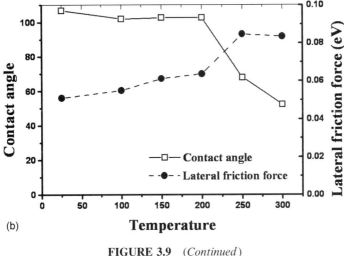

(b)

FIGURE 3.9 (*Continued*)

Figures 3.10 and 3.11 show the water contact angle and lateral friction force as functions of UV exposure time at room temperature. Figure 3.10 indicates that the effectiveness of the SAM deposition declined for UV exposure times greater than 450 s. Figure 3.11 indicates that the lateral friction force tended to decrease with increasing exposure time, similar to the experiments with maximum replication temperature. These results show that the molecular binding of SAM with low surface energy is damaged by longer

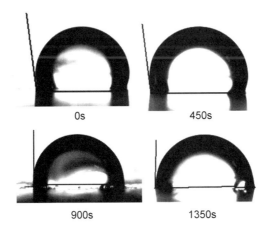

FIGURE 3.10 Images of water contact angle for different UV-exposure time at room temperature.

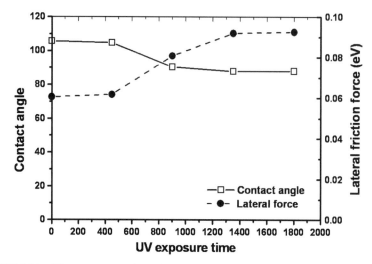

FIGURE 3.11   Water contact angle and lateral friction force as function of UV-exposure time at room temperature.

UV exposure times to the point in actual UV imprinting where the anti-adhesion property of an SAM-deposited mold may deteriorate due to UV overexposure. However, the contact angle was stable at about 90° after 1350 s of UV exposure. The contact angle of the SAM-deposited nickel mold after UV exposure was higher than that after the maximum replication temperature, indicating that the damage of the SAM due to UV exposure was less than that due to heat energy. A comparison of the contact angle after UV exposure of the bare nickel mold and the SAM-deposited molds in Table 3.2 shows that the SAM-deposited molds were more hydrophobic after UV exposure. High-energy UV light is absorbed by UV curing polymers under actual UV imprinting conditions. Therefore, an SAM-deposited mold can be expected to maintain its anti-adhesion properties for more than 450 s under actual process conditions.

3.2.4   Analysis of Replicated Polymeric Patterns

Figure 3.12 shows the AFM images of nickel mold and replicated polymeric patterns by hot-embossing process, where the embossing temperature was 190°C. Figure 3.12a shows the AFM image and surface profile of nickel mold. Figure 3.12b and c shows the AFM image and surface profile of the embossed patterns; with the bare nickel mold and SAM-deposited nickel mold, respectively. In the case of Figure 3.12b, replicated patterns were deformed due to sticking problem and had poor surface quality of 30.3 Å rms; the mold had the surface roughness of the 13.2 Å. While, in the case of Figure 3.12c, the replicated patterns by the SAM-deposited mold were

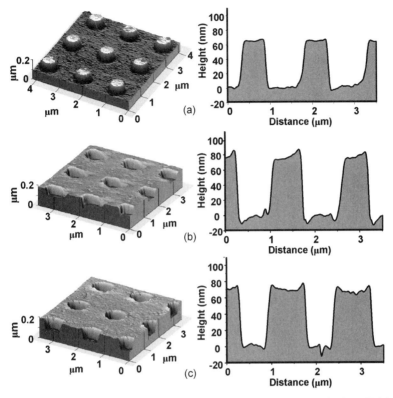

FIGURE 3.12 (a) AFM image and surface profile of the nickel mold. (b) AFM image and surface profile of replicated parts replicated by the bare nickel mold. (c) AFM image and surface profile of replicated parts replicated by the SAM-deposited nickel mold. Reprinted with permission from Ref. [9]. Copyright 2006, American Institute of Physics.

replicated well and had good surface quality of 16.2 Å rms roughness. When comparing the surface profiles of replicated patterns with those of mold pattern as shown in Figure 3.12, the surface profile of Figure 3.12c is more similar to those of mold pattern in Figure 3.12a. These results verified the effectiveness of n-dodecanethiol SAM as an anti-adhesion layer for the actual-replication process with nickel mold [15].

To apply our method to replication of sub-100 nm scale nanopillar arrays, the nanopatterned substrate with nanopillar arrays was replicated using nanoinjection molding process with SAM-deposited nickel mold. It is difficult to replicate nanopillar arrays with sub-100 nm using the conventional replication process without anti-adhesion layer due to the sticking problem between the mold and the replicated patterns. Figure 3.13 shows SEM images of the nanopatterned substrate replicated by nanoinjection molding process with SAM-deposited nickel mold. The nanopatterned substrate had pillar patterns with a diameter of 100 nm, a pitch of 250 nm, a height of 100 nm, and

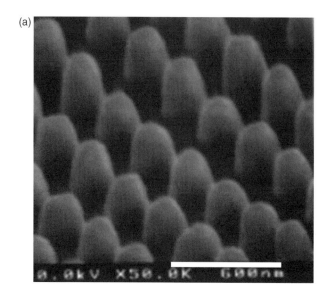

FIGURE 3.13 SEM image of nanopillar arrays with (a) 100 nm diameter, 250 nm pitches and (b) with 50 nm diameter, 150 nm pitches, which were replicated by nanoinjection molding process with the SAM-deposited nickel mold. Reprinted with permission from Ref. [9]. Copyright 2006, American Institute of Physics.

a diameter of 50 nm, a pitch of 150 nm, a height of 50 nm, as shown respectively in Figure 3.13. These results verified that sub-100 nm scale nanopillar arrays with good surface quality can be replicated by nanoinjection molding process with SAM-deposited nickel mold.

3.3 SILANE-BASED SELF-ASSEMBLED MONOLAYER

3.3.1 Silane-Based Self-Assembled Monolayer

Many defects, such as sticking, occur due to the adhesion force between the molds and polymer during UV nanoimprinting. However, the removal of curable polymer is difficult and risks damage to the mold because the curable polymer is very stable. To increase the releasing properties of the silicon mold, an SAM of tridecafluoro-1,1,2,2-tetrahydrooctyltrichlorosilane $(CF_3(CF_2)_5(CH_2)_2SiCl_3$, FOTS) was coated onto the mold surface as an anti-adhesion layer [16–17].

Figure 3.14a shows the composition of tridecafluoro-1,1,2,2-tetrahydrooctyltrichlorosilane monomer with $SiCl_3^-$ of the monomer combining with the substrate forming the reaction group, and an alkyl group from the thickness of the monolayer with a function group making up the hydrophobic surface. As shown in Figure 3.14b, the monomers combine with the substrate in an orderly manner to form a very thin and stable monolayer. A series of silane monolayer depositions on the substrate can be produced primarily by two methods: the solution method and the vaporization method. The solution method is a relatively simple process and requires only simple equipment, but

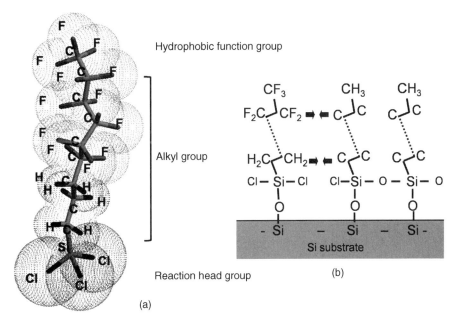

FIGURE 3.14 (a) Tridecafluoro-1,1,2,2-tetrahydrooctyltrichlorosilane $(CF_3(CF_2)_5(CH_2)_2$ $SiCl_3$, FOTS). (b) Schematic structure of FOTS monolayers on the silicon substrate. Reprinted with permission from Ref. [9]. Copyright 2006, American Institute of Physics.

may lead to problems of pattern damage during the pretreatment and surface degradation due to adverse reactions. In contrast, the vaporization method requires more complicated processes and equipment. A uniform monolayer is deposited using the vaporization process and has excellent surface properties as an anti-adhesion layer.

3.3.2 Deposition Process of Silane-Based Self-Assembled Monolayer

The vaporization method is composed of two steps: pretreatment and reaction. Pretreatment is required to modify the substrate surface with O_2 plasma so that it becomes hydrophilic to induce the reaction of SAM molecules with the substrate. Because the Piranha process, another pretreatment method, carries with it the risk of damaging the nanopatterns, the O_2 plasma method is more suitable for nanopattern molds. Many types of SAM deposition equipment have only a single chamber in which the O_2 plasma process and the SAM reaction process take place. However, many SAM molecules remain in the chamber after repeated process cycles. Residual molecules damaged by the O_2 plasma of subsequent processes are deposited irregularly, as shown in Figure 3.15. These damaged molecules on the surface interrupt the reaction between the SAM precursor and the substrate, and the uniformity of the SAM deteriorates to the point where the expected properties are not achieved. An SAM deposition system with two chambers was developed, as shown in Figure 3.16 to prevent the SAM molecules from being damaged by the plasma. Figure 3.17 shows contact angle results of FOTS SAM deposited on a silicon mold in single- and dual-chamber systems. Figure 3.17a shows that the contact angles of a specimen deposited in the single-chamber system were in the range 92.7–100.2°. On the other hand, Figure 3.17b shows that the contact angles of a specimen deposited in the dual-chamber system were stable at about 108°. This indicates that the mold produced in the dual-chamber system is more hydrophobic and stable than one produced in the single-chamber system.

3.3.3 Self-Assembled Monolayer on Polymer Mold

Because the bare surface of an organic–inorganic mold does not have good release properties, an anti-adhesion SAM was coated on the mold surface. A silane anti-adhesion layer cannot be coated on normal organic material because the coating process is very harsh. The target substrate must be heated to 200–220°C after being coated with the SAM material. In addition, hydrochloric acid is formed as a by-product of the SAM coating. No normal organic material can survive such conditions. However, hybrid

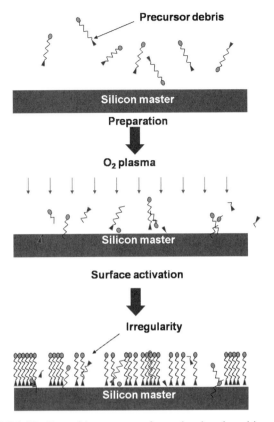

FIGURE 3.15   Deposition process of one-chamber deposition system.

organic–inorganic photopolymer is highly reliable in the face of such high temperatures and acid chemical solvents. The weight loss of this photopolymer is less that 5% for temperatures up to 270°C, and hydrochloric acid will not dissolve it. The contact angle of the bare hard mold surface was compared with that of the SAM-coated hard mold surface. The contact angle of the hard

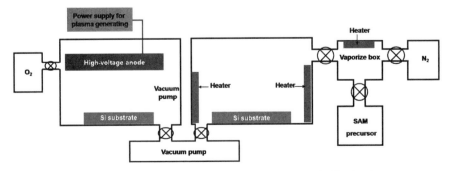

FIGURE 3.16   Chamber system for deposition of anti-adhesion layer.

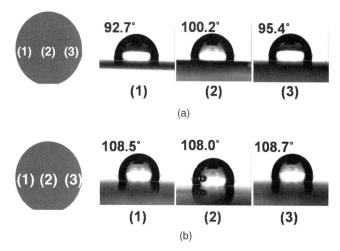

FIGURE 3.17 (a) Contact angle of SAM on silicon master using one chamber deposition system, (b) contact angle of SAM on silicon master using two chamber deposition system. Reprinted with permission from Ref. [17]. Copyright 2011, American Scientific Publishers.

mold was measured by the same method described above. The average contact angle of the bare hard mold surface was 38.6°, which is inadequate for UV imprinting. However, the surface of the SAM-coated hard mold was hydrophobic. Figure 3.18 and Table 3.3 show the results of the contact angle measurement. The fabrication process based on this result is described in more detail in Chapter 6.

3.3.4 Analysis of Replicated Polymeric Patterns

UV nanoimprinting using a silicon mold makes release of the replica from the mold difficult due to the adhesion force between the polymer and mold. Figure 3.19a shows the unexpected defects of UV nanoimprinting using a silicon mold without an anti-adhesion layer. To avoid these defects in the replica, FOTS SAM was deposited on the silicon mold with nanopatterns using the method described earlier. Figure 3.19b shows the result of UV nanoimprinting with an SAM-coated silicon mold.

3.4 DIMETHYLDICHLOROSILANE SELF-ASSEMBLED MONOLAYER

Nano replication using a silicon mold is very risky due to the brittleness of silicon. Techniques using metal molds such as nickel have been developed to overcome this. However, molds made of metals such as nickel have many inherent problems with replication due to their adhesion properties.

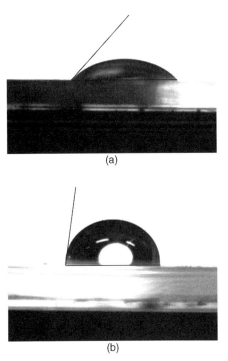

(a)

(b)

FIGURE 3.18 Photopolymer drop on hard mold surface (a) bare hard mold surface and (b) SAM-coated mold surface.

The anti-adhesion layer described in this chapter can solve these problems due to adhesion force of interfaces. In the case of silane SAM, described in Section 3.2, hydrochloric acid from the reaction between the SAM precursor and the substrate causes the corrosion of metals or metal oxides. Because of problems such as this, the deposition of silane SAM on metal surfaces has been studied in some detail. This section describes coating a dimethyldichlorosilane (DDMS) SAM on a nickel mold.

The molecular formula of DDMS is $(CH_3)_2SiCl_2$. It forms a very small monomer that is a thinner monolayer than FOTS due to its lower molecular

TABLE 3.3 Result of the Contact Angle of Hard Mold

	Contact Angle (Bare Surface)	Contact Angle (SAM-Coated Surface)
Sample 1	38.6°	71.1°
Sample 2	37.5°	70.1°
Sample 3	39.6°	69.9°
Average	38.6°	70.4°

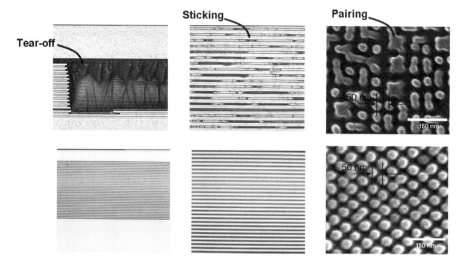

FIGURE 3.19 Fabrication results by UV nanoimprinting (a) without anti-adhesion layer on silicon master and (b) with anti-adhesion layer on silicon master. Reprinted with permission from Ref. [17]. Copyright 2011, American Scientific Publishers.

weight [18]. FOTS with three chlorine atoms is highly reactive and produces more hydrochloric acid after the reaction. However, DDMS with two chlorine atoms is less reactive. This difference of reactivity results in a low probability of polymerization of the SAM precursor in the SAM coating processes.

The process of coating a nickel mold with DDMS is shown in Figure 3.21. After washing and rinsing the nickel mold, a surface oxidation layer is formed for DDMS to react with the substrate. When deposited on the metal surface, the oxidation layer coating is capable of preventing corrosion by the hydrochloric acid. SiO_2 is deposited using plasma-enhanced chemical vapor deposition. DDMS is deposited on the metal surface using a deeping

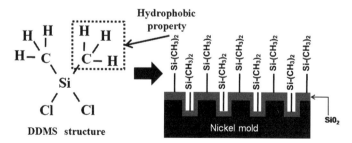

FIGURE 3.20 Schematic diagram of chemical structure of dichlorodimethylsilane (DDMS; $Cl_2Si(CH_3)_2$) on the nickel mold.

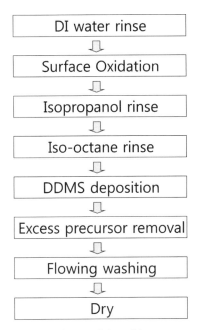

FIGURE 3.21 Flowchart of DDMS deposition process; solution method.

process, and then the excess precursor on the surface is removed using an isooctane rinse.

After the deposition of DDMS, the adhesion property of the DDMS SAM was analyzed by contact angle measurements. Figure 3.22 shows contact angle images of a bare nickel mold and a nickel mold with a DDMS SAM. The contact angle increased from 60° to 114° after the SAM deposition. The SiO$_2$ deposited to act as a precursor with the nickel mold was so thin (about 20 nm) that the nickel mold could be used in UV nanoimprinting to replicate micropatterns. UV roll nanoimprinting with this mold is described in Chapter 6.

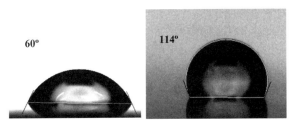

FIGURE 3.22 (a) Water contact angle on bare nickel mold. (b) Water contact angle on DDMS–SAM-deposited nickel mold.

REFERENCES

1. N. Lee, Y. Kim, and S. Kang (2004) Temperature dependence of anti-adhesion between a stamper with sub-micron patterns and the polymer in nano-moulding processes. *Journal of Physics D: Applied Physics* 37, 1624–1629.
2. N. Lee, Y. Kim, S. Kang, and J. Hong (2004) Fabrication of metallic nano-stamper and replication of nano-patterned substrate for patterned media. *Nanotechnology* 15, 901–906.
3. N. Lee, S. Moon, S. Kang, and S. Ahn (2003) The effect of wettability of nickel mold insert on the surface quality of molded microlenses. *Optical Review* 10, 290–294.
4. K. Seong, S. Moon, and S. Kang (2001) An optimum design of replication process to improve optical and geometrical properties in DVD-RAM substrates. *Journal of Information Storage and Processing Systems* 3, 169–176.
5. R. W. Jaszewski, H. Schift, B. Schnyder, A. Schneuwly, and P. Gröning (1999) The deposition of anti-adhesive ultra-thin teflon-like films and their interaction with polymers during hot embossing. *Applied Surface Science* 143, 301–308.
6. M. Bender, M. Otto, B. Hadam, B. Spangenberg, and H. Kurz (2002) Multiple imprinting in UV-based nanoimprint lithography-related material issues. *Microelectronic Engineering 61–62* 407–413.
7. O. Azzaroni, P. L. Schilardi, and R. C. Salvarezza (2003) Metal electrodeposition on self-assembled monolayers: a versatile tool for pattern transfer on metal thin films. *Electrochimica Acta* 48, 3107–3114.
8. Y. Xia, X.-M. Zhao, and G. M. Whitesidesz (1996) Pattern transfer: self-assembled monolayers as ultrathin resists. *Microelectronic Engineering* 32, 255–268.
9. N. Lee, S. Choi, and S. Kang (2006) Self-assembled monolayer as an anti-adhesion layer on a nickel nanostamper in the nanoreplication process for optoelectronic applications. *Applied Physics Letters* 88(7), 073101-1–073101-3.
10. H. I. Kim, V. Boiadjiev, J. E. Houston, X.-Y. Zhu, and J. D. Kiely (2001) Tribological properties of self-assembled monolayers on Au, SiOx and Si surfaces. *Tribology Letters* 10, 97–101.
11. F. Laffineur, J. Delhalle, and Z. Mekhalif (2002) Surface coating of a Cu–Ni alloy with a self-assembled monolayer of n-dodecanethiol. *Material Science and Engineering C* 22, 331–337.
12. B. J. Kim, M. Liebau, J. Huskens, D. N. Reinhoudt, and J. Brugger (2001) A self-assembled monolayer-assisted surface microfabrication and release technique. *Microelectronic Engineering 57–58* 755–760.
13. Z. Mekahalif, J. Riga, J.-J. Pireaux, and J. Delhalle, (1997) Self-assembled monolayers of *n*-dodecanethiol on electrochemically modified polycrystalline nickel surfaces. *Langumir* 13, 2285–2290.
14. Z. Mekhalif, F. Laffineur, N. Couturier, and J. Delhalle (2003) Elaboration of self-assembled monolayers of *n*-alkanethiols on nickel polycrystalline substrates: time, concentration, and solvent effects. *Langmuir* 19, 637–645.

15. N. Lee, S. Choi, and S. Knag (2006) Self-assembled monolayer as an anti-adhesion layer on a nickel nanostamper in the nanoreplication process for optoelectronic applications. *Applied Physics Letters* 88, 073101.

16. M. Bender, M. Otto, B. Hadam, B. Spangenberg, and H. Kurz (2002) Multiple imprinting in UV-based nanoimprint lithography-related material issues. *Microelectronic Engineering 61–62* 407–413.

17. S. Choi, J. Han, J. Lim, H. Kim, S. Kim, and S. Kang (2011) Development of a two-chamber process for self-assembling a fluorooctatrichlorosilane monolayer for the nanoimprinting of full-track nanopatterns with a 35 nm half pitch. *Journal of nanoscience and nanotechnology* 11, 5921–5927.

18. B. Kim, T. Chung, C. Oh, and K. Chun (2001) A new organic modifier for anti-stiction. *Journal of Microelectromechanical System* 10, 33–40.

CHAPTER 4

Micro/Nanoinjection Molding with an Intelligent Mold System

4.1 INTRODUCTION

In injection molding, raw plastic is melted inside a barrel and injected under high pressure into a mold cavity, where it cools until it retains the shape of the cavity. The molded part is then removed from the cavity [1]. Since injection molding offers high productivity, repeatable high tolerances, a fully automated process, and near-net-shape products, and is applicable to a wide range of materials, the process is extensively used to manufacture various plastic optical, biomedical, and structural components. Thanks to the increasing demand for plastic products with micro/nanopatterns in the fields of optical data storage, optical communications, digital display, and digital imaging, as well as for energy, fluidic, and biomedical devices, the micro/nanoinjection molding technique is becoming a priority. Figure 4.1 shows some examples of commercially available micro/nanoinjection-molded parts, including (a) optical data storage (ODD) media (BD), (b) a light-guided plate (LGP) for a liquid crystal display (LCD) backlight unit (BLU), (c) a micro-Fresnel lens, and (d) a microfluidic channel.

The development of microinjection molding began in the late 1980s, using a hydraulic injection molding system with clamping forces of 25–50 t [2]. Since the volume of the micropart and the required injection force are only a fraction of the normal injection volume and injection pressure, a substantial amount of material loss occurred, and fine control of the injection molding processing parameters was impossible. Furthermore, to obtain high-fidelity micro/nanostructures on the injection-molded parts, more injection pressure was required than was readily available. To overcome these problems, a micro/nanoinjection molding system that offers small injection volumes and fine control of injection pressure has been developed for micro/nanoproducts, such as micropumps and

Micro/Nano Replication: Processes and Applications, First Edition. Shinill Kang.
© 2012 John Wiley & Sons, Inc. Published 2012 by John Wiley & Sons, Inc.

(a)

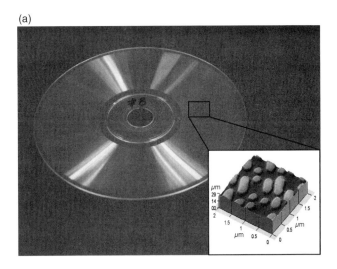

(b)

FIGURE 4.1   Images of various injection molded micro/nanoplastic product: (a) optical data storage media; (b) LGP for LCD BLU; (c) micro-Fresnel lens; (d) microchannels.

microgears [3,4]. An injection compression molding system, in which the injection of the polymer melt into the mold cavity is followed by a compression process in order to reduce the injection pressure and increase the pattern fidelity, has also been developed for thin-plate products with micro/nanopatterns, such as ODD media and LGPs for LCD backlights [5,6]. In addition, a mold system has been developed for micro/nanoinjection molding. To fabricate a micro/nanomold cavity structure in the mold system, mechanical machining of a metal block, known as a mold insert, is generally used. The electroforming of a micro/nanopatterned substrate, called a stamper, is also widely employed.

(c)

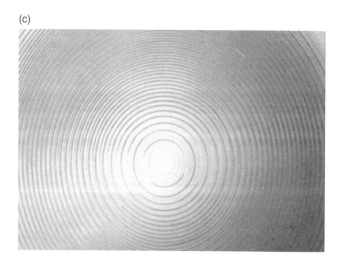

(d)

FIGURE 4.1   (*Continued*)

Along with the development of the micro/nanoinjection molding system, the molding process has also been developed [7–10]. To replicate micro/nanoparts with high fidelity, specific processing conditions must be chosen. Various studies on the effects of processing conditions on the micro/nanoinjection of molded parts have shown that the main process parameters affecting the products are (1) mold temperature, (2) injection speed, (3) injection pressure, (4) holding time, and (5) holding pressure. One of the critical issues in the micro/nanoinjection molding process is unfilled micro/nanocavities due to the solidification layer generated on the surface of the mold by the temperature difference between the polymer melt and the mold surface. In this chapter, a theoretical basis for the generation of the solidification layer, as well as methods of preventing this layer from forming in optical data storage media by using an intelligent mold system, will be discussed.

4.2 EFFECTS OF THE MOLD SURFACE TEMPERATURE ON MICRO/NANOINJECTION MOLDING

Replication quality is the most important factor in the micro/nanoinjection molding process. In the past few years, numerous experimental and theoretical investigations have been undertaken to develop process technologies for micro/nanoinjection molding. Mönkkönen et al. [11] experimentally analyzed the effects of various process conditions on the micro/nano replication process. Kang et al. [12] carried out an experimental investigation of the effects of the processing parameters on the shape of injection-molded microchannels, and optimized the parameters. Seong et al. [13] studied the effects of the process parameters on the principal replica properties in optical disc substrate molding, and proposed a process optimization technique based on the relationship between critical process parameters and principal replica properties. Schift et al. [14] analyzed injection molding and hot-embossing processes for replicating nanostructures.

Among the various process parameters of the injection molding process, the surface temperature of the mold is critical for micro/nanoinjection molding, since the micro/nanopatterns on the replicas and the thicknesses of the products have both decreased in size. Figures 4.2 and 4.3 illustrate the effects of the mold surface temperature on the replication quality of injection-molded nanopatterns with different substrate thicknesses. Figure 4.2 shows atomic force microscopy (AFM) images of the mold and molded patterns for high-density optical discs: (a) the pit nanopattern for read-only memory (ROM) optical disks and (b) the land-groove nanopattern for rewritable (RW) optical disks. Figure 4.3 shows the heights of the pit and land-groove nanopatterns of the replicated substrates for various mold temperatures using (a) type I substrate with a thickness of 1.1 mm and an outer diameter of 83 mm and (b) type II substrate with a thickness of 0.4 mm and an outer diameter of 47 nm. The patterns in the mold have a track pitch of 0.32 μm, a minimum pit length of 0.16 μm, and a pit height of 850 Å. The samples were fabricated with an electric actuation injection molding machine with a clamping force of 40 t, and a high-precision mold temperature controller with a control accuracy of $\pm 0.1°C$, depicted in Figure 4.4. The filling time and packing pressure were set to 0.1 s and 400 kgf/cm^2, respectively. To raise the resin temperature to 300°C, the temperature distribution in the cylinder was divided into five

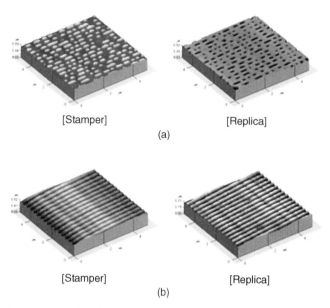

[Stamper]   [Replica]

(a)

[Stamper]   [Replica]

(b)

FIGURE 4.2 Stamper and replica for high-density optical disc with pitch of 0.32 μm: (a) pit; (b) land-groove structure (mold temperature: 110°C). Reprinted with permission from Ref. [15]. Copyright 2005, Springerlink.

sections. From the section nearest to the sprue, the temperature distribution was 300, 320, 360, 320, and 300°C. The mold temperature was set at 90, 100, 110, and 120°C to analyze the effect of the stamper surface temperature. For the type I optical disk with pit structures (ROM) shown in Figure 4.3a, the height of the molded patterns increased from 650–760 Å as the mold temperature increased from 90–120°C. In the land-groove structures (RW), patterns with heights of 260, 325, and 605 Å were replicated at mold temperatures of 90, 100, and 120°C, respectively. For the type II optical disk shown in Figure 4.3b, the pit height increased with mold temperature, but at a mold temperature of 120°C, the heights were much lower than those of the type I optical disk. These results show that transcribability can be improved by increasing the mold temperature, and deteriorates with decreasing substrate thickness. Also note that the ROM nanopattern was much more easily filled by the polymer melt than the RW nanopattern.

During the filling stage, the polymer melt rapidly solidifies in the vicinity of the mold surface, and the solidified layer generated during polymer filling greatly impairs the transcribability and fluidity of the melt. As Figure 4.3 indicates, this problem is more critical for injection-molded thin products with micro/nanopatterns, such as ODD media and LGPs. Yoshii et al. [16] investigated the effect of the solidified layer on the micro/nano replication process, using a mathematical model. Kim et al. [17] numerically analyzed

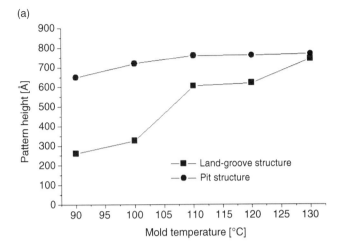

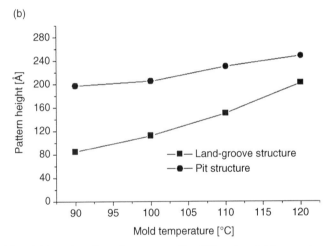

FIGURE 4.3 Height of replicated structures for different mold temperature: (a) type I (substrate thickness of 1.1 mm and outer diameter of 83 mm); (b) type II (substrate thickness of 0.4 mm and outer diameter of 47 mm). Reprinted with permission from Ref. [15]. Copyright 2005, Springerlink.

the effects of the mold surface temperature on the growth of the solidified layer and the viscosity of the polymer melt. Figure 4.5 illustrates how the height of a molded pattern decreases as the thickness of the solidified layer increases. Here, d_1 and d_2 denote the thickness of the solidified layer and the height of the molded pattern, respectively. To improve the transcribability and fluidity of the polymer melt, the mold surface can be heated above the glass transition temperature of the polymer in order to delay the growth of the solidified layer during the filling stage. However, there are many technical

FIGURE 4.4   Injection molding system for micro/nanomolding process.

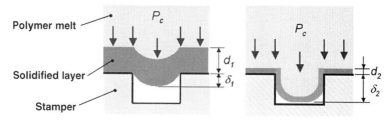

FIGURE 4.5   Deterioration of transcribability due to solidified layer in the micro/nanoinjection molding process.

problems associated with using conventional heating methods to accomplish this, due to the high thermal mass of the mold. Heating the entire mold results in a longer cycle time. Therefore, it is advantageous to control only the surface temperature of the mold or the stamper.

4.3   THEORETICAL ANALYSIS OF PASSIVE/ACTIVE HEATING METHODS FOR CONTROLLING THE MOLD SURFACE TEMPERATURE

In the conventional control method for mold temperature, the entire mold, which has a large thermal inertia, is heated and cooled to maintain a fixed mold temperature. Therefore, the mold surface temperature is difficult to control during the filling process when conventional injection molding is used. Various techniques have been adopted to heat the mold surface during the filling stage. Kim and Suh [18], Liou and Suh [19], and Kim and Niemeyer [20] suggested an insulated mold insert to retain the heat in the mold cavity; this is known as passive heating. Passive heating is a simple and effective approach. Inoue et al. [21] and Kim et al. [22] replicated a high-density optical disc substrate using an insulation layer. However, passive heating using an insulation layer or insulated mold insert presents some critical issues. The first of these is cycle time. When passive heating is applied to the replication process, the mold surface temperature is effectively increased, but so is the cooling time, and thus the cycle time is also increased. The second issue is replication quality. When the mold surface temperature is increased by passive heating, the transcribability of the replica is improved. However, the improvement of the transcribability is limited, since the mold surface temperature is increased due to the retardation of heat transfer by the insulation layer, as depicted in Figure 4.6a.

To avoid these passive heating issues, an active heating technique has been proposed. In active heating, the cycle time can be shortened and the mold surface temperature raised since the mold surface is directly heated by a heater, as depicted in Figure 4.6b. For micro/nano replication, an active heating system has been developed that does not simply heat the mold surface, but also controls the mold surface temperature according to the optimum temperature history. This is important because the temperature

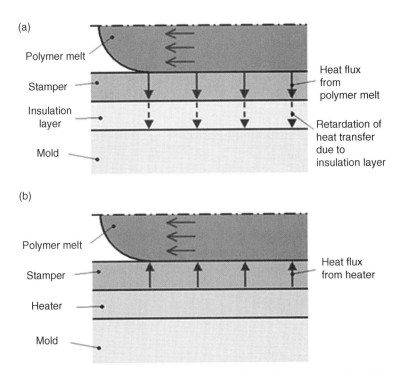

FIGURE 4.6 Method for heating the stamper surface: (a) passive heating; (b) active heating.

history of the mold surface has a critical effect on the optical and mechanical properties of the replica [23].

In the next section, a flow analysis will be introduced into the theoretical analysis of melt flow in injection molding. The results for passive heating and active heating will be discussed in sections 4.3.2 and 4.3.3.

4.3.1 Mathematical Modeling and Simulation

For a flow analysis of injection molding with passive heating, a mathematical model is constructed in the cavity (Ω_m) and in the mold, consisting of the stamper $(\partial\Omega_{st})$, the first insulation layer(Ω_{ins1}), a heating layer (Ω_h), the second insulation layer (Ω_{ins2}), and a mold block (Ω_{mb}), as shown in Figure 4.6. In passive heating, the heating layer and the second insulation layer are omitted.

In this section, a mathematical model is developed for a flow analysis in the cavity of an optical disk substrate with axisymmetric radial flow of a polymer melt. The mass, momentum, and energy equations, based on the Hele-Shaw approximation, are as follows [24]:

$$\frac{\partial\rho}{\partial t} + \frac{1}{r}\frac{\partial\rho ru}{\partial r} = 0, \quad r \in \Omega_m \tag{4.1}$$

$$-\frac{\partial p}{\partial r} + \frac{\partial}{\partial z}\left(\eta\frac{\partial u}{\partial z}\right) = 0, \quad r, z \in \Omega_m \tag{4.2}$$

$$-\frac{\partial p}{\partial z} + \frac{\partial}{\partial z}\left(\eta\frac{\partial w}{\partial z}\right) = 0, \quad z \in \Omega_m \tag{4.3}$$

$$\rho C_p\left(\frac{\partial T}{\partial t} + u\frac{\partial T}{\partial r}\right) = \frac{\partial}{\partial z}\left(k\frac{\partial T}{\partial z}\right) + \eta\dot{\gamma}^2, \quad z \in \Omega_m \tag{4.4}$$

where, ρ, T, t, u, and w are the density, temperature, time, and velocities in the r and z directions, respectively, k denotes thermal conductivity, Cp denotes specific heat, η is the viscosity coefficient, and $\dot{\gamma}$ represents the shear rate. For axisymmetric radial flow,

$$\dot{\gamma} = \left|\frac{\partial u}{\partial z}\right| \tag{4.5}$$

The formula for the viscosity is obtained from the p–v–T behavior and the Cross/WLF model [25]:

$$\eta = \eta_0(p,\, T)\left[\frac{1 + s\left(\frac{\eta_0 \dot{\gamma}}{\tau^*}\right)^{1-n}}{1 + \left(\frac{\eta_0 \dot{\gamma}}{\tau^*}\right)^{1-n}}\right] \tag{4.6}$$

where η_0 is given as follows:

$$\eta_0 = D_1 \exp\left[-\frac{A_1(T-T_g)}{A_2 + (T-T_g)}\right], \quad T \geqslant T_g \tag{4.7}$$

$$\eta_0 = D_1 \exp\left[-\left(\frac{A_1 T_g}{A_2}\right)\left(\frac{T-T_g}{T}\right)\right], \quad T < T_g \tag{4.8}$$

$$A_2 = \tilde{A}_2 + D_3 p, \quad T_g = D_2 + D_3 p \tag{4.9}$$

T_g is the glass transition temperature, and n, τ^*, D_1, D_2, D_3, A_1, $\tilde{A}_2$, and s are property variables with respect to the materials. Using the double-domain Tait model [25] for the p–v–T behavior, the following relationships between the density, pressure, and temperature are obtained:

$$\rho(T,\, p) = \rho_0(T)\left\{1 - C\ln\left[1 + \frac{p}{B(T)}\right]\right\}^{-1} \tag{4.10}$$

$$\frac{1}{\rho_0(T)} = \begin{cases} b_{1,l} + b_{2,l}(T-b_5) & \text{for } T > T_t \\ b_{1,s} + b_{2,s}(T+b_5) & \text{for } T \leq T_t \end{cases} \tag{4.11}$$

$$B(T) = \begin{cases} b_{3,\,l}\exp(-b_{4,l}T) & \text{for } T > T_t \\ b_{3,\,s}\exp(-b_{4,s}T) & \text{for } T \leq T_t \end{cases} \tag{4.12}$$

$$T_t(p) = b_5 + b_6 p \tag{4.13}$$

where $b_{1,l}$, $b_{2,l}$, $b_{3,l}$, $b_{4,l}$, $b_{1,s}$ $b_{2,s}$, $b_{3,s}$, $b_{4,s}$, b_5, and b_6 are property variables with respect to the materials. The coefficients for the Cross/WLF model [26] and the double-domain Tait [25] equation are listed in Tables 4.1 and 4.2.

TABLE 4.1   Cross/WLF Model Coefficients of Polycarbonate with Reference Temperature 287°C

Coefficients	Values
n	0.170
τ^* [Pa]	6.91×10^5
D_1 [Pa-s]	5.82×10^9
D_2 [°C]	144
D_3 [°C/Pa]	1.9×10^{-7}
A_1	22.8
$\tilde{A}_2\,°[C]$	61.2
s	1.0×10^{-5}

For thermal analysis in a mold with a heating layer, the transient one-dimensional heat conduction equation can be used:

$$\rho C_p \left(\frac{\partial T}{\partial t} \right) = \frac{\partial}{\partial z} \left(k \frac{\partial T}{\partial z} \right) + S \qquad (4.14)$$

where S is the heating term corresponding to the heat source.

- Active heating:

$$S = \frac{\dot{W}}{dV} \qquad (4.15)$$

- Passive heating:

$$S = 0 \qquad (4.16)$$

TABLE 4.2   Double-Domain Tait Equation Coefficients of Polycarbonate

Coefficients	Values
$b_{1,1}$ [cm^3/g]	9.799×10^{-1}
$b_{2,1}$ [cm^3/(g°C)]	5.788×10^{-4}
$b_{3,1}$ [dyne/cm^2]	1.483×10^9
$b_{4,1}$ [1/°C]	3.019×10^{-3}
$b_{1,s}$ [cm^3/g]	9.799×10^{-1}
$b_{2,s}$ [cm^3/(g°C)]	2.429×10^{-4}
$b_{3,s}$ [dyne/cm^2]	1.965×10^9
$b_{4,s}$ [cm^3/g]	1.380×10^{-3}
b_5 [°C]	1.03×10^2
b_6 [(°C cm^2)/dyne]	3.2×10^{-8}

where $\dot{W}$ is the heat generated in the heating layer and V is the volume of the heating layer. The initial temperature of the polymer melt is set as a uniform temperature T_m. The temperatures of the stamper, insulation layer, heating layer, and mold are initially set to the mold wall temperature T_w.

The boundary conditions of the numerical model are as follows: at the mold surface,

$$T = T_w \text{ at } Z = H \tag{4.17}$$

and in the cavity,

$$u = 0 \text{ at } z = h \tag{4.18}$$

$$\frac{\partial w}{\partial z} = \frac{\partial T}{\partial z} = 0 \text{ at } z = 0 \tag{4.19}$$

$$\text{and } p = 0 \text{ at the melt front } (\partial \Omega_{mf}). \tag{4.20}$$

For the flow analysis in the cavity, a finite element and finite difference hybrid scheme can be used. A finite element method (FEM) is used to solve the flow fields calculated by Eqs. (4.1), (4.2), and (4.3). A finite difference method (FDM) is used to solve the temperature fields in the cavity calculated by Eq. (4.4), and an FDM [27] with a control volume formulation [28] is used to solve the temperature fields in the mold calculated by Eq. (4.14).

The flow analysis in the cavity and thermal analysis in the injection mold are carried out simultaneously by using the equilibrium heat flux boundary condition at the stamper surface, namely the interface between the polymer melt and the stamper. For the flow analysis in the cavity, an initial temperature estimate $T_i(r,h,t)$ is used as the boundary temperature for the equilibrium heat flux boundary condition at the stamper surface. The thermal analysis in the mold is carried out by using the heat flux $q_i(r,h,t)$ calculated in the flow analysis as the boundary condition at the stamper surface. The temperature $T_{st}(h,t)$ at the stamper surface is then calculated in the mold. The thermal and flow analyses are repeated until the convergence condition at the stamper surface defined by Eq. (4.21) is satisfied. This procedure is followed until the polymer filling is completed.

$$|T_i(r, h, t) - T_{st}(h, t)| < \text{Tolerance at the stamper surface } (\partial \Omega_{st}) \tag{4.21}$$

A cooling analysis at the stamper surface is also carried out via the same technique. Figure 4.7 illustrates the numerical analysis procedure for the flow analysis of injection molding with a heating layer.

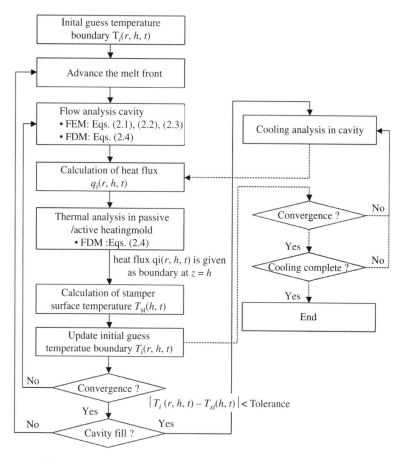

FIGURE 4.7   Numerical analysis procedure for flow analysis of injection molding with active heating. Reprinted with permission from Ref. [17]. Copyright 2004, Institute of physics science.

4.3.2   Passive Heating

In this section, the effects of passive heating on the stamper surface temperature, the development of the solidified layer, and the transcribability of a molded optical disk are discussed.

Figure 4.8 shows the temperature history of the stamper surface for different insulation layer thicknesses (passive heating) after the polymer melt contacts the stamper surface. The mold cavity for the simulation had a thickness of 0.6 mm, which is typical for a high-density optical disk RW substrate. The initial melt temperature and mold temperatures were 320 and 90°C, respectively. Polycarbonate was used as the polymer material, and nickel was used for the stamper and heating layer. Tool steel was chosen for

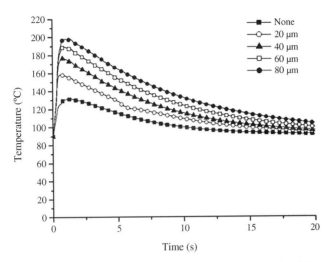

FIGURE 4.8 Simulated temperature history of the stamper surface for different insulation layer thicknesses. Reprinted with permission from Ref. [17]. Copyright 2004, Institute of physics science.

the mold material, and polyimide was chosen for the insulation layer. The thermal properties of each material are listed in Table 4.3.

The maximum stamper surface temperature increased with increasing insulation thickness. For insulation thicknesses greater than $40\,\mu m$, the stamper surface temperature remained above the glass transition temperature ($144°C$) for several seconds, since an insulation layer with low inertia retards the heat flow from the polymer melt to the steel mold, causing the insulation layer to reheat the stamper surface. This implies that the temperature difference between the polymer melt and the stamper surface decreases effectively during injection molding. Also, the stamper surface temperature decreases to around $100°C$ in 20 s, regardless of the insulation thickness. Consequently, the temperature history of the stamper surface shows that the stamper surface temperature can be elevated and maintained above the glass transition temperature during filling.

TABLE 4.3 Thermal Properties of Layers

Layer	$\rho[kg/m^3]$	$C_p[J\,kg^{-1\circ}C^{-1}]$	$k[W\,m^{-1\circ}C^{-1}]$
Polymer	1017	2232	0.294
Stamper	8880	460	60.7
Insulation layer	1610	1092	0.155
Heating layer	8880	460	60.7
Mold	7800	460	26

To examine the development of the solidified layer and the effect of passive heating on its development, polymer melt flow with passive heating was analyzed. The flow rate, melt temperature, and packing pressure were 11.88 cm³/s, 320°C, and 13.8 MPa, respectively. To predict the development of a solidified layer, the glass transition temperature given by the p–v–T equation of state was used [29]. The thickness of the solidified front, D_s, is given as follows:

$$D_s = h - \left\{ z_i + \left(\frac{T_i - T_g}{T_i - T_{i-1}} \right) \Delta Z \right\}$$ (4.22)

where h, z_i, and T_i are the half-thickness of the cavity, the z-coordinate at the ith node, and the temperature at the ith node, respectively. T_g is the glass transition temperature.

Figure 4.9 shows the profile of the solidified front with respect to time during polymer filling, when the polymer melt fills the cavity for 0.617 s. When there was no insulation (Figure 4.9a), a solidified front developed with time. However, for insulation thicknesses of 40 and 80 µm (shown in Figure 4.9b and c, respectively), the initial solidified front vanished with respect to time, and a new solidified front developed. This is caused by the increased stamper surface temperature due to passive heating, and the increased polymer melt temperature resulting from dissipation heating [29]. As a result, the solidified layer melted again during polymer filling. This behavior indicates that passive heating can markedly delay the development of the solidified layer during polymer filling. There was very little difference between the thicknesses of the solidified fronts in Figures 4b and c, because for insulation thicknesses greater than 40 µm, the stamper surface temperature remained above the glass transition temperature during polymer filling, as shown in Figure 4.8.

To verify these predictions, a high-density optical disk RW substrate was fabricated by injection molding with passive heating [30]. An injection molding machine with a clamping force of 35 t was used. Optical-grade polycarbonate (Lexan OQ 1020c-112), a common material for optical disk substrates, was used to mold a substrate with an outer diameter of 120 mm and a thickness of 0.6 mm. An insulation layer made of polyimide was inserted under the stamper. Figure 4.10 shows the injection mold with the stamper and thermal insulation layer. Figure 4.11 shows AFM images and surface profiles of the land-groove structure of the injection-molded optical disk. The disk was fabricated at a mold temperature of 90°C (a) without an insulation layer, and with insulation layer thicknesses of (b) 33 µm, and (c) 78 µm. The groove depth and track pitch of the stamper were 500 Å and 0.68 µm, respectively. The 90°C mold temperature did not provide complete transcribability, as shown in Figure 4.11a. However, Figures 4.11b and c indicate that passive heating

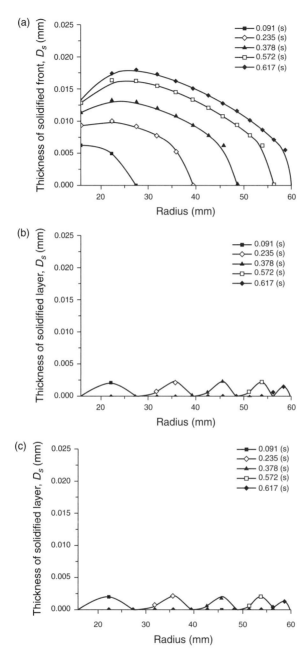

FIGURE 4.9 Advancement of solidified front with respect to time: (a) without insulation layer; (b) with insulation thickness of 40 μm; (c) with insulation thickness of 80 μm (fill time = 0.617 s). Reprinted with permission from Ref. [17]. Copyright 2004, Institute of physics science.

FIGURE 4.10   Injection mould of horizontal type with a stamper and a thermal insulation layer for fabricating optical disk substrates. Reprinted with permission from Ref. [22]. Copyright 2002, SPIE.

dramatically improved the transcribability, independent of the insulation thickness. This implies that transcribability can be improved by delaying the development of the solidified layer and increasing the fluidity of the polymer melt during polymer filling in the vicinity of the stamper surface.

4.3.3   Active Heating

In this section, the effects of active heating on the surface temperature of the mold and the viscosity and fluidity of the polymer melt will be discussed. A method to predict the optimal surface temperature for obtaining sufficient fluidity of the polymer will also be introduced. For the simulation, a mold cavity with a thickness of 0.4 mm and a diameter of 47 mm was used. The initial melt temperature and mold temperature were 320 and 100°C, respectively. Polycarbonate, nickel, and polyimide were chosen for the polymer material, stamper and heating layer, and insulation layer, respectively. The thermal properties of these materials are listed in Table 4.3. In active heating, the maximum stamper surface temperature and heating rate are increased as the power density is increased, as shown in Figure 4.12. At a power density of 50 W/cm², the stamper surface temperature increased from 100–270°C in 1 s. Also, the stamper surface was cooled down to the mold temperature (100°C) within a few seconds, which did not significantly increase the cycle time.

The solidified layers that occurred with active heating, passive heating, and conventional injection molding were simulated. Figure 4.13 shows the

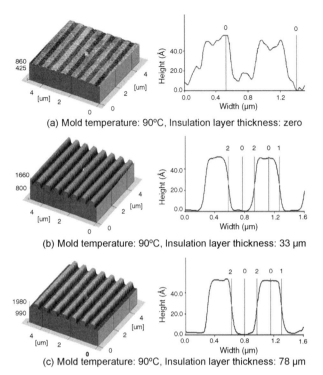

(a) Mold temperature: 90°C, Insulation layer thickness: zero

(b) Mold temperature: 90°C, Insulation layer thickness: 33 μm

(c) Mold temperature: 90°C, Insulation layer thickness: 78 μm

FIGURE 4.11   AFM images (left) and surface profiles (right) of the land-groove structure at different insulation layer thickness: (a) mould temperature, 90°C; insulation layer thickness, zero; (b) mould temperature, 90°C; insulation layer thickness, 33 μm; (c) mould temperature, 90°C; insulation layer thickness, 78 μm. Reprinted with permission from Ref. [22]. Copyright 2002, SPIE.

simulation results. In active heating, a power density of $20\,\text{W/cm}^2$ was used for 1 s. In passive heating, an insulation thickness of $40\,\mu\text{m}$ was used. The simulation results suggest that the thickness of the solidified layer is effectively decreased by active heating, and this implies that transcribability in the micro/nano replication process can be more significantly improved by active heating than by either passive heating or conventional injection molding. Also, when the polymer melt viscosity decreases, active heating improves its fluidity more than passive heating or conventional injection molding. Figure 4.14a shows the simulation results for the viscosity distribution in the direction of substrate thickness. $z = 0$ and 0.2 mm are the center of the cavity and the stamper surface, respectively. The viscosity, which attained its peak value near the stamper surface, decreased from 9.14×1013–$4.52 \times 1010\,\text{gm/cm·s}$, compared with conventional injection molding. An increase in fluidity due to a decrease in viscosity will facilitate the micro/nano replication process. The residual stress on the replica can be

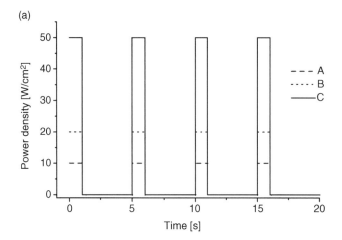

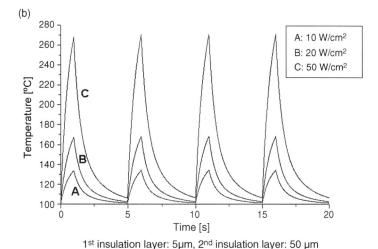

1st insulation layer: 5μm, 2nd insulation layer: 50 μm

FIGURE 4.12 Heating of stamper surface by active heating (simulation results): (a) input power density and (b) stamper surface temperature. Reprinted with permission from Ref. [15]. Copyright 2005, Springerlink.

reduced as the fluidity is increased or the viscosity is decreased. Figure 4.14b shows the simulation results for fluidity.

To improve the replication quality in micro/nanoinjection molding, the viscosity of the polymer melt must be decreased and the fluidity must be increased by increasing the mold surface temperature. Since an excessive increase in surface temperature may cause unexpected problems, it is important to select the optimal surface temperature via analysis of viscosity surface response. To analyze the effects of the mold surface temperature on viscosity, we calculated the viscosity under various process conditions.

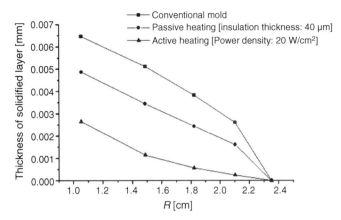

FIGURE 4.13   Solidified layer at the end of filling.

Polycarbonate was used as the material. Equation (4.6), together with the coefficients listed in Table 4.1, was used to calculate the viscosity. First, the viscosity surface at a pressure of 30 MPa was predicted, where 30 MPa is the simulated average pressure in the cavity for an optical disc substrate with a diameter of 47 mm and a thickness of 0.4 mm at the conclusion of filling. Figure 4.15 shows the results. The viscosity rapidly increased for a stamper surface temperature greater than 180°C and a shear rate greater than 2000/s. This implies that fluidity can be effectively improved by raising the stamper surface temperature above 180°C.

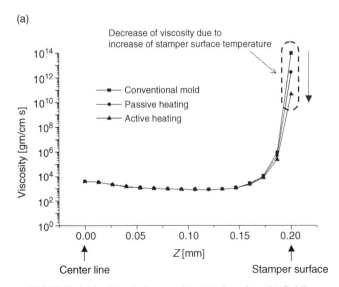

FIGURE 4.14   Simulation results: (a) viscosity; (b) fluidity.

(b)

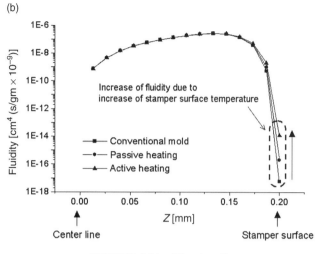

FIGURE 4.14 (*Continued*)

4.4 FABRICATION AND CONTROL OF AN ACTIVE HEATING SYSTEM USING AN MEMS HEATER AND AN RTD SENSOR

To realize active heating, the development of a mold system with a heater and sensor and an efficient controller system are required. This intelligent mold system should increase and decrease the stamper surface temperature by about 100°C within a few seconds. Various heaters have been proposed to actively heat the stamper surface. Jansen and Flaman [31] used carbon resin as a heating element. Yao and Kim [32] heated the stamper surface by supplying electric

(a)

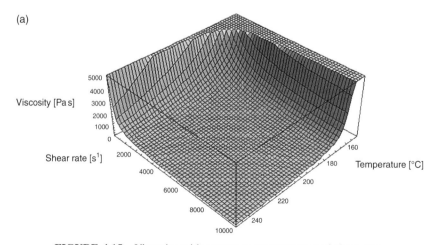

FIGURE 4.15 Viscosity with respect to temperature and shear rate.

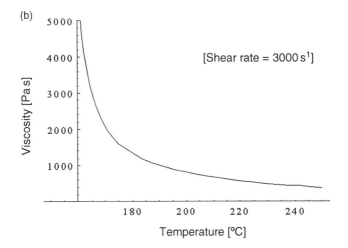

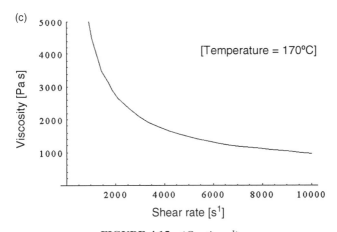

FIGURE 4.15   (*Continued*)

power to a metal layer. Saito et al. [24] adopted an infrared laser to actively heat the stamper surface.

In this section, the fabrication and control of an intelligent mold system using a microelectromechanical system (MEMS) heater and sensor will be discussed in detail. Such a system has relatively small thermal inertia, uses single lithographic techniques to integrate the heater and sensor, and is considered the most suitable method for micro/nanoinjection molding.

4.4.1 Construction of an Intelligent Mold System

The intelligent mold system described in this section is designed to replicate an optical disc substrate with an outer diameter of 47 mm, and has a double-sided stamper structure. A thin-film heater is used to raise the stamper surface temperature above the glass transition temperature during the filling stage. The thin-film heater is constructed on the backside of the stamper.

A temperature sensor must be built into the injection mold to monitor the stamper surface temperature. Conventional thermocouples are typically employed for this purpose. However, thermocouples do not satisfy the processing speed requirements of injection molding because of their slow response, nonrepeatability, and large size [23]. Luo et al. developed thin-film sensors for online monitoring of injection molding [33–35].

For an intelligent mold system, a thin-film resistance temperature detector (RTD) sensor, which can be simultaneously fabricated with an MEMS heater, is preferable. To measure the stamper surface temperature effectively, the RTD sensor must be constructed on the stamper surface or mold wall. Figure 4.16a shows the fabrication process for the thin-film MEMS heater and RTD sensor.

A polyimide film was spin-coated on the stamper as the first insulation layer. Polyimide is a good insulation material because it has high thermal

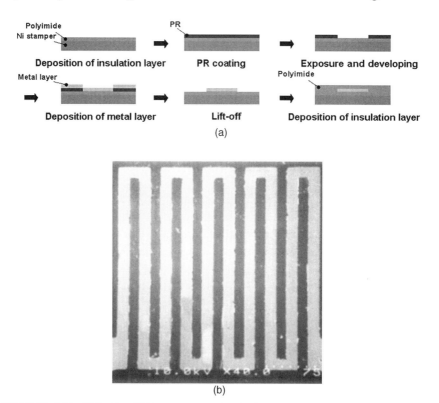

(a)

(b)

FIGURE 4.16 Thin-film RTD sensor. (a) fabrication process; (b) SEM image of the fabricated sensor. Reprinted with permission from Ref. [15]. Copyright 2005, Springerlink.

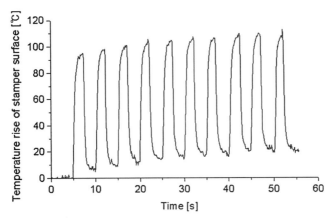

FIGURE 4.17 Temperature rise of stamper surface by thin-film heater. Reprinted with permission from Ref. [15]. Copyright 2005, Springerlink.

stability, low dielectric constant, high electrical resistivity, and excellent adhesion. Next, a metal layer was fabricated on the polyimide film using a lithography and lift-off process. Nickel was used in this study for the thin metal layer. Nickel and platinum have been widely used in the fabrication of RTD sensors because their electric resistances vary linearly over a wide range of temperatures. To protect the deposited sensor from moisture, corrosion, and damage, another layer of polyimide was spin-coated on the sensor to form the passivation layer and second insulation layer. Figure 4.16b shows a scanning electron microscopy (SEM) image of the fabricated sensor.

Figure 4.17 shows the measured stamper temperatures for an applied input power density; these were used to study the characteristics of the fabricated MEMS heater. The stamper temperature could be raised by about 100°C within a few seconds. Thus, the present thin-film heater can heat the stamper surface above the glass transition temperature of polycarbonate during the filling stage, since the conventional mold temperature for an optical disc is about 100°C.

The RTD sensor was annealed for 60 min in the heating chamber at various temperatures to thermally stabilize the metal thin film and to obtain a better understanding of the annealing effects. After annealing, the sensor was cooled down to room temperature, and its resistance was measured. As Figure 4.18 indicates, the resistance of the nickel thin film decreased as the annealing temperature increased. It is believed that the nanovoids were coalesced by grain growth during the annealing process. Figure 4.19 shows that the nanovoids in the deposited metal layer diminished as the annealing temperature increased. An annealing temperature of 300°C was chosen for

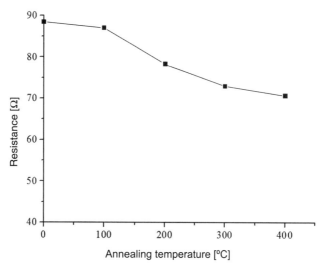

FIGURE 4.18   Effects of annealing temperature on the resistance of deposited nickel thin film.

the thin-film RTD sensor because the stamper surface temperature did not exceed 300°C during the injection molding process.

A thin-film RTD sensor measures the transient temperature by determining the resistance change of the metal film. Therefore, the resistance and temperature characteristics of the nickel thin film should be determined before the actual measurements are taken. The resistance of the nickel film with respect to temperature was measured by using a signal-conditioning module in the heating chamber, together with a temperature controller and a thermocouple.

(a)

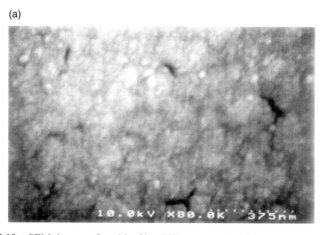

FIGURE 4.19   SEM images for thin-film RTD sensor: (a) before annealing; (b) after annealing.

(b)

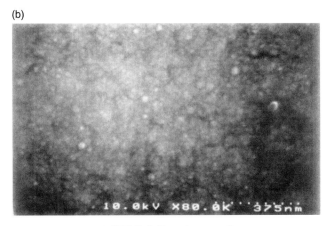

FIGURE 4.19   *(Continued)*

The four-wire connection method was used in the signal-conditioning module to eliminate the effects of the lead wire resistance. Figure 4.20 shows the experimental setup for measuring the resistance of the nickel thin film with respect to temperature. Figure 4.21 shows the calibration results for the thin-film RTD sensor from 0–300°C. The experimental results indicate that the resistance is nonlinear with respect to the temperature, but repeatable.

The experimental data were correlated using the following equation:

$$R = R_o(1 + AT + BT^2 + CT^3) \qquad (4.22)$$

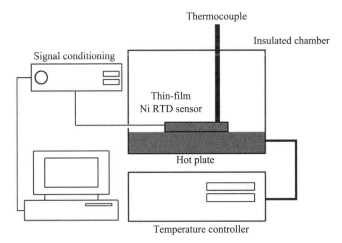

FIGURE 4.20   Experimental setup for measuring resistance of thin-film RTD sensor with respect to temperature.

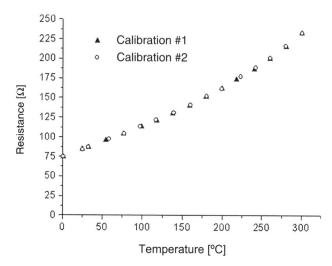

FIGURE 4.21   Calibration results for thin-film RTD sensor. Reprinted with permission from Ref. [15]. Copyright 2005, Springerlink.

where R and T denote the base resistance at $0°C$ and the temperature, respectively, and A, B, and C are constants related to a given RTD sensor. From the experimental results, the following equation can be obtained:

$$R = 74.5667 \left(1 + 5.2148 \times 10^{-3}\,T - 2.0390 \times 10^{-6}\,T^2 + 2.7102 \times 10^{-8}T^3\right)$$

$$(4.23)$$

4.4.2 Control System for the Intelligent Mold System

In the micro/nano replication process, active heating of the stamper surface requires a system that does not simply heat the stamper surface, but also controls the stamper surface temperature in accordance with the optimum temperature history, since the temperature history of the stamper surface affects the optical and mechanical properties of the replica [36]. Control schemes for the injection molding process have previously been researched. Tan *et al.* controlled the ram velocity of an injection molding machine with a PI control [37]. Gao et al. applied an iterative learning control (ILC) to an injection molding machine to control the injection velocity [38]. However, in these studies, the control scheme was incorporated into the machine itself.

To construct a control system for active heating, the control module should be integrated with the injection mold, along with the MEMS heater and MEMS resistance temperature detector sensor described in the previous section. To control the stamper surface temperature, a control system for active heating of

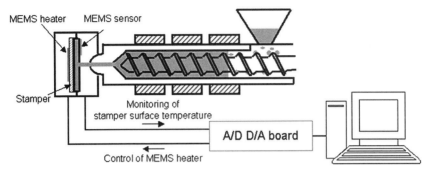

FIGURE 4.22 Closed-loop control system for active heating of stamper surface. Reprinted with permission from Ref. [15]. Copyright 2005, Springerlink.

the stamper surface was designed and constructed, as depicted in Figure 4.22. The stamper surface temperature was monitored by a thin-film RTD sensor, and the optimum input for the MEMS heater was based on the measured temperature. As a result, the stamper surface temperature was controlled by the thin-film heater. A closed-loop controller was designed, using a Kalman filter and a linear quadratic Gaussian regulator. The stamper surface temperature was then controlled by using the injection mold with the MEMS heater and MEMS RTD sensor that was fabricated in the previous chapter.

A plant model consisting of an injection mold and an MEMS heater will be described as depicted in Figure 4.23. A multi-layer structure composed of a

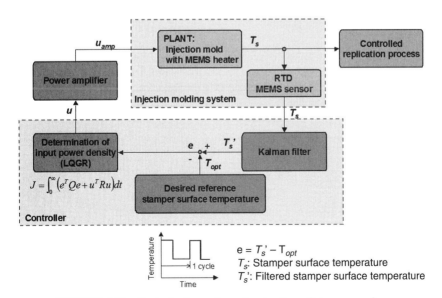

FIGURE 4.23 Control scheme for active heating of stamper surface.

stamper (Ω_{st}), a first insulation layer (Ω_{ins1}), a heating layer (Ω_h), a second insulation layer (Ω_{ins2}), and a mold (Ω_m) were fabricated. Since heat conduction in the thickness direction (z-direction) is dominant in the thermal model, the transient one-dimensional heat conduction equation can be applied,

$$\rho_i c_i \frac{\partial T}{\partial t} = k_i \frac{\partial^2 T}{\partial z^2} + S_i \tag{4.24}$$

where k_i, ρ_i, and c_i are the thermal conductivity, density, and specific heat, respectively. Also, i represents a given layer: the stamper, the first insulation layer, the heating layer, the second insulation layer, or the mold. The heat source associated with the MEMS heater is denoted by S_i for all layers ($\forall i \in \Omega$), and can be obtained from the following:

$$S_i = \begin{cases} \dfrac{\dot{W}}{\Delta V} & if \ i \in \Omega_h \\ 0 & if \ i \in \Omega \backslash \Omega_h \end{cases} \tag{4.25}$$

where $\dot{W}$ [W] is the heat generated in the heating layer and ΔV is the volume of the heating layer. The principal thermal properties of each layer are listed in Table 4.3. In order to obtain the discrete-time state-space model of the thermal plant directly, the FDM was used to discretize Eq. (4.24) for the thermal plant [27]. The boundary conditions of the model consisted of a heat convection condition at the stamper surface, and a heat flux condition at the end of mold.

Let us define the state vector of the thermal plant by the following:

$$x_p(k) = [x_1(k), x_2(k), \cdots, x_N(k)]^T \in R_{11} \tag{4.26}$$

where $x_i(k) \equiv T_i^k$ is the temperature of a given layer at the discrete time k. The discrete-time state-space representation for the thermal plant model with node temperature state $x \in R_{11}$ can be obtained directly by combining the previous equations,

$$A_p^{-1} x(k+1) = x(k) + N_1 U + N_2 \tag{4.27}$$

where the input vector $U \in R_2$ is given by the following:

$$U = \begin{bmatrix} \dot{W} & T_{mold} \end{bmatrix}^T \in R_2 \tag{4.28}$$

Based on the model obtained in the previous section, a linear quadratic Gaussian with integrator (LQGI) controller, similar to the one used in

Ref. [39], was designed and implemented to solve the stamper surface temperature-tracking problem. The design of the Kalman filter observer and the LQGI controller are discussed in the following sections.

4.4.2.1 Kalman Filter Observer of the Thermal Plant   The discrete-time state-space representation of the thermal plant is given by the following:

$$x_p(k+1) = A_p x(k) + B_{p1} u_f(k) + B_{p2} t_{mold} + B_{p2} w_m(k) + B_{p3} t_e + B_{p3} w_e(k)$$

$$y(k) = C_p x_p + v(k) \tag{4.29}$$

where $A_p \in R_{11 \times 11}$, $B_{p1} \in R_{11 \times 1}$, $B_{p2} \in R_{11 \times 1}$, $B_{p3} \in R_{11 \times 1}$, $C_p \in R_{1 \times 11}$, and $u_f \in R$, are the input power density from the amplifier, $v(k)$ is the white noise from the temperature sensor, $t_{mold} \in R$ is the controlled fixed mold temperature, and $w_m \in R$ is the white noise of the fluctuation from the room temperature. Thus, the mold temperature T_{mold} and the room temperature T_e can be expressed by the following:

$$T_{mold}(k) = t_{mold} + w(k)$$

$$T_e(k) = t_e + w_e(k) \tag{4.30}$$

The three white noises $v(k)$, $w_m(k)$, and $w_e(k)$ are statistically independent each other, and have the following properties.

The amplifier generating the power density u_f is the input–output system empirically identified by the prediction error estimate method, using the system identification toolbox in Matlab:

$$x_f(k+1) = A_f x_f(k) + B_f u(k) + K_f e(k)$$
$$u_f(k) = C_f x(k) + D_f u(k) + e(k) \tag{4.31}$$

where

$$A_f = \begin{bmatrix} 0.97355 & 0.058099 & -0.069526 \\ -0.001177 & 0.28781 & 0.64266 \\ -0.029605 & -0.55414 & 0.16957 \end{bmatrix}$$

$$B_f = \begin{bmatrix} 0.0014096 \\ -0.013638 \\ 0.010557 \end{bmatrix}$$

$$C_f = [\,176.3 \quad -74.067 \quad 106.82\,]$$

$$D_f = 0$$

$$K_f = \begin{bmatrix} 0.00060881 \\ 0.00059769 \\ 0.00067307 \end{bmatrix}$$

and $e(k)$ is the Gaussian white noise satisfying the following:

$$E(e(k)) = 0, \; E(e(i)e(j)) = \delta_{ij} \qquad (4.32)$$

The state of the augmented system $x_a = \begin{bmatrix} x_p^T & x_f^T \end{bmatrix}^T \in R_{14}$ consists of the thermal plant state $x_p \in R_{11}$ and the amplifier state $x_f \in R_3$. The augmented system is defined by the following:

$$\begin{aligned} x_a(k+1) &= A_a x_a(k) + B_a u_a(k) + G_a w(k) \\ y_a(k) &= C_a x_a + v(k) \end{aligned} \qquad (4.33)$$

where the system matrices are $A_a \in R_{14 \times 14}$, $B_a \in R_{14 \times 3}$, $G_a \in R_{14 \times 3}$, and $C_a \in R_{1 \times 14}$, and $w(k)$ and $v(k)$ are white noise.

Given the detectable pair (A_a, C_a) and the stabilizable pair (A_a, B_a) of the augmented system defined by Eq. (4.33), there exists a steady-state Kalman filter gain, which is given by the following:

$$F_s = \Sigma_s C_a^T (C_a \Sigma_s C_a^T + V)^{-1} \qquad (4.34)$$

where Σ_s is the solution of the algebraic Riccati equation:

$$\Sigma_s = A_a \Sigma_s A_a^T + G_a W G_a^T - A_a \Sigma_s C_a^T (C_a \Sigma_s C_a^T + V)^{-1} C_a \Sigma_s A_a^T \qquad (4.35)$$

Using the steady-state Kalman filter gain obtained from Eq. (4.34) and the augmented system of Eq. (4.33), the optimal state estimator can be constructed as follows:

$$\begin{aligned} \hat{x}_a(k+1|k+1) \\ = (I - F_s C_a) A_a \hat{x}(k|k) + (I - F_s C_a) B_a u_a(k) + F_s y_a(k+1) \end{aligned} \qquad (4.36)$$

where $\hat{x}_a(k|k)$ represents an a posteriori estimate for $x_a(k)$ or the optimal estimate for $x_a(k)$, based on the observations up to time k.

4.4.2.2 LQGI Controller An integral action control is implemented to eliminate the constant disturbances in the temperature-controlled system.

Let $e(k) = y_a(k) - y_r(k)$ denote the tracking error between the reference trajectory $y_r(k) = T_{st}(k)$ of the desired stamper surface temperature $T_{st}(k)$ and the measured stamper surface temperature $y_a(k)$. The incremental tracking error integration $I(k)$ is given by the following:

$$\begin{aligned} I(k+1) &= I(k) + e(k) \\ &= I(k) + C_a x_a(k) - y_r(k) \end{aligned} \tag{4.37}$$

For greater freedom in shaping the reference trajectory $y_r(k)$, a second-order linear dynamic system with a unit static gain was chosen for the reference generating model:

$$\begin{aligned} x_r(k+1) &= A_{r(2\times 2)} x_r(k) \\ &\quad + B_{r(2\times 1)} x_r(k) + B_{r(2\times 1)} u_r(k) \\ y_r(k) &= C_{r(1\times 2)} x_r(k) \end{aligned} \tag{4.38}$$

We define an extended state x^{track} by

$$x^{track}(k) = \begin{bmatrix} x_a(k) \\ I(k) \\ x_r(k) \end{bmatrix} \in R17 \times 1 \tag{4.39}$$

Consider the composite system consisting of the augmented thermal system, the integrator dynamics, and the reference model as follows:

$$\begin{aligned} x^{track}(k+1) &= A^{track} x^{track}(k) \equiv \\ &\begin{bmatrix} A_{a(14\times 14)} & 0_{14\times 1} & 0_{14\times 2} \\ C_{a(1\times 14)} & 1 & -C_{r(1\times 2)} \\ 0_{2\times 14} & 0_{2\times 1} & A_{r(2\times 2)} \end{bmatrix} x^{track}(k) \\ &+ \begin{bmatrix} B_{a1(14\times 1)} & B_{a2(14\times 1)} & B_{a3(14\times 1)} & 0_{15\times 1} \\ 0_{3\times 1} & 0_{3\times 1} & 0_{3\times 1} & B_{r(2\times 1)} \end{bmatrix} \\ &\begin{bmatrix} u(k) \\ t_{mold} \\ t_e \\ u_r(k) \end{bmatrix} \end{aligned}$$

$$y^{track}(k) = (y_a(k) - y_r(k)) + gI(k)$$
$$y^{track}(k) = C^{track}x^{track} = [C_a \quad g \quad -C_r]x^{track}(k)$$

$$(4.40)$$

where B_{ai} is the ith column vector of B_a,

$$B_a = \begin{bmatrix} B_{a1(14\times1)} & B_{a2(14\times1)} & B_{a3(14\times1)} \end{bmatrix}$$

and g is the weighting factor between the tracking error and the incremental tracking error. Let us define the backward-shift operator q^{-1} such that $q^{-1}h(k) = h(k-1)$ for all discrete sequences $h(k)$.

We define a new set of state variables x_d^{track}, x_d, and x_{rd} by the following:

$$x_d^{track} \equiv \begin{bmatrix} x_d(k) \\ e(k) \\ x_{rd}(k) \end{bmatrix} \equiv (1-q^{-1})x^{track}(k+1) = \begin{bmatrix} (1-q^{-1})x_a(k+1) \\ (1-q^{-1})I(k+1) \\ (1-q^{-1})x_r(k+1) \end{bmatrix}$$

$$(4.41)$$

By multiplying each side of Eq. (4.42) by the difference operator $(1 - q^{-1})$, and taking into account that t_{mold}, t_e, and $u_r(k)$ are constant,

$$\begin{bmatrix} x_d(k) \\ e(k) \\ x_{rd}(k) \end{bmatrix} = A^{track}\begin{bmatrix} x_d(k-1) \\ e(k-1) \\ x_{rd}(k-1) \end{bmatrix} + \underbrace{\begin{bmatrix} B_{a1(14\times1)} \\ 0_{3\times1} \end{bmatrix}}_{B^{track}} v(k-1) \qquad (4.42)$$

where $v(k)$ is defined by $v(k-1) = (1-q^{-1})u(k)$. Equation (4.42) can be written as follows:

$$x_d^{track}(k) = A^{track}x_d^{track}(k-1) + B^{track}v(k-1)$$
$$y_d^{track}(k-1) = C^{track}x_d^{track}(k-1)$$

$$(4.43)$$

Consider a quadratic tracking performance index given by the following:

$$J = \sum_{k_0}^{\infty} \left(\begin{array}{c} x^{track^T}(k)Q^{track}x_d^{track}(k) \\ +v(k)^T R^{track}v(k) \end{array} \right)$$

$$(4.44)$$

where $Q^{track} = C^{track^T}C^{track}$ and $R^{track} > 0$ is a penalty weighting factor for the incremental control input $v(k)$. The optimal quadratic control, with penalties on the tracking error and the incremental control input, is given by the following:

$$
\begin{aligned}
u(k) &= -K^{track} x^{track}(k)\\
&= -K_a x_a(k) - K_r x_r(k) - K_e \sum_{k_0}^{k-1} e(k)
\end{aligned}
\tag{4.45}
$$

where the optimal control gain matrix is given by the following:

$$
\begin{aligned}
K^{track} &= \begin{bmatrix} K_{a(1\times14)} & K_{e(1\times1)} & K_{r(1\times2)} \end{bmatrix}\\
&= (R^{track} + B^{track} P B^{track})^{-1} B^{track^T} P A^{track}
\end{aligned}
$$

and P is the solution of the following algebraic Riccati equation:

$$
\begin{aligned}
&A^{track^T} P A^{track} - P -\\
&A^{track^T} P B^{track} K^{track} + Q^{track} = 0
\end{aligned}
\tag{4.46}
$$

4.4.2.3 *Performance of the Constructed Control System*   The desired reference stamper surface temperature is determined to calculate the error between the target temperature and the measured temperature. The cycle time and the heating duration are 5 and 1 s, respectively. The maximum reference temperature and the mold temperature are 200 and 100°C, as depicted in Figure 4.24. The stamper surface is heated from 100–200°C just after the onset of the filling stage. The stamper surface temperature is

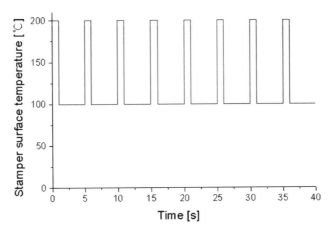

FIGURE 4.24   Desired reference stamper surface temperature.

maintained at the maximum reference temperature of 200°C for 1 s, and is then held at the mold temperature of 100°C through the remaining cycle time.

To analyze the effect of the desired reference stamper surface temperature on the micro/nano replication process, flow analysis of the polymer melt was carried out at the reference temperature. Figure 4.25 shows the simulation results for the solidified layer and the viscosity. A solidified layer did not develop during the filling stage at the reference stamper surface temperature. Also, compared with conventional injection molding, the viscosity in the vicinity of the stamper surface was drastically decreased, from 2.34×10^5–1.32×10^{13} gm/cm·s. These simulation results indicate that the

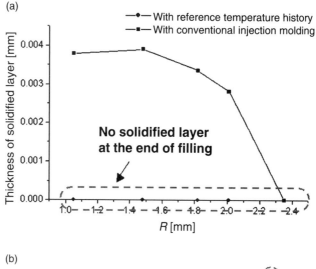

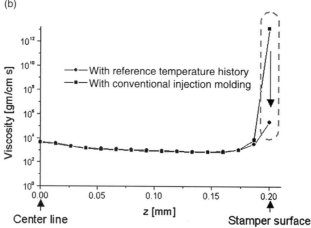

FIGURE 4.25 Simulation results with desired reference stamper surface temperature: (a) solidified layer; (b) viscosity.

quality of micro/nano replication can be improved at the desired reference stamper surface temperature.

An experimental setup was developed to test the proposed control scheme. Using an NI PCI-6024E as an I/O board, together with a personal computer, a data I/O system was constructed. A control algorithm with a sampling time of 0.01 s was built in Simulink. A power amplifier with a maximum output power of 150 W was used to amplify the signal from the I/O board. To analyze the system characteristics and verify the control scheme, three known inputs were first loaded to the MEMS heater. The first input was a sine wave whose frequency varied linearly with time. The frequency of the first input changed from 0.01–10 Hz in 200 s. Figure 4.26a shows the system response for the first input. The temperature profile simulated by the designed plant model was similar to the measured temperature. The second and third inputs were colored noises. As Figures 4.26b and c indicate, the simulated temperature profiles were similar to the measured temperature profiles. After the verification of the plant model, the gain values of g and R were adjusted to tune the controller.

Figure 4.27 shows the control results for the stamper surface temperature at temperature increments of 100°C. These results indicate that the stamper

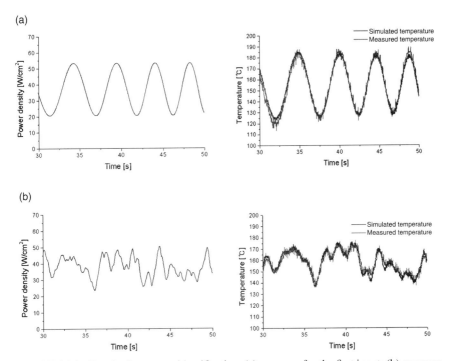

FIGURE 4.26   Results for system identification: (a) response for the first input; (b) response for the second input; (c) Response for the third input.

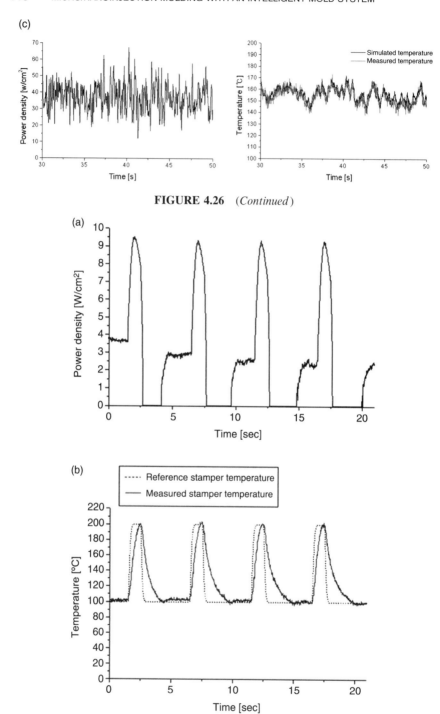

FIGURE 4.26 (*Continued*)

FIGURE 4.27 Control results for stamper surface temperature (temperature increment: 100°C): (a) input power density; (b) measured temperature profile.

surface temperature can be controlled in accordance with a desired reference temperature profile. However, the cooling rate was overestimated.

4.5 REPLICATION OF A HIGH-DENSITY OPTICAL DISC SUBSTRATE USING THE INTELLIGENT MOLD SYSTEM

In this section, the feasibility of the intelligent mold system is demonstrated on the HD optical disk described in section 4.1. Figure 4.28 compares the injection-molded nanopatterns with and without the intelligent mold system. The intelligent mold system was used to replicate an HD optical disk with a thickness of 0.4 mm and diameter of 47 mm (type II), and the pattern height of the conventional injection-molded parts was ~ 200 Å. Since the height of the structures on the stamper was 850 Å, to guarantee more than 90%

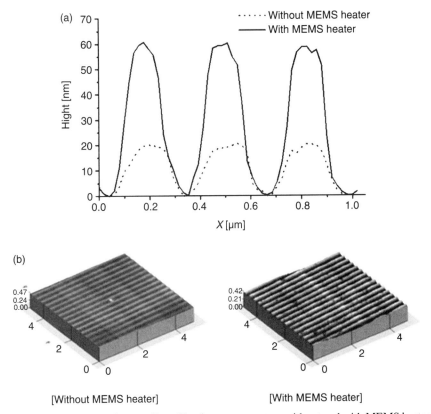

FIGURE 4.28 (a) Surface profiles of land-groove structure without and with MEMS heater; (b) Molded land-groove structure without and with MEMS heater (track pitch: 0.32 μm). Reprinted with permission from Ref. [15]. Copyright 2005, Springerlink.

transcribability, the mold surface was heated above the glass transition temperature of polycarbonate (145°C) to delay the growth of the solidified layer during the filling stage. However, there are many technical problems associated with raising the mold temperature above the glass transition temperature of polycarbonate via conventional methods. Because of its high thermal mass, it is not easy to heat the entire mold to a temperature greater than 145°C using conventional temperature controllers. Such heating would result in a longer cycle time. Therefore, it is advantageous to control only the surface temperature of the mold or the stamper. By using the MEMS heater at a mold temperature of 110°C, the height of a molded land-groove structure with a pitch of 0.32 μm increased from about 202–607 Å, which demonstrates the effectiveness of the intelligent mold system [40].

REFERENCES

1. M. P. Groover (2007) *Injection Molding in Fundamentals of Modern Manufacturing*, Wiley, New York, 275, 286.
2. V. Piotter, K. Mueller, K. Plewa, R. Ruprecht, and J. Hausselt (2002) Performance and simulation of thermoplastic microinjection molding. *Microsystem Technologies* 8, 387–390.
3. J. Giboz, T. Copponnex, and P. Mele (2007) Microinjection molding of thermoplastic polymers: a review. *Journal of Micromechanics and Microengineering* 17, R96–R109.
4. W. Michaeli, A. Spennemann, and R. Gartner (2002) New plastification concepts for microinjection moulding. *Microsystem Technologies* 8, 55–57.
5. M. Heckele and W. K. Schomburg (2004) Review on micromolding of thermoplastic polymers. *Journal of Micromechanics and Microengineering* 14, R1–R14.
6. S. C. Chen, Y. C. Chen, N. T. Cheng, and M. S. Huang (1998) Simulation of injection-compression mold-filling process. *International Communications in Heat and Mass Transfer* 25, 907–917.
7. B. R. Whiteside, M. T. Martyn, and P. D. Coates (2005) In process monitoring of micromoulding – assessment of process variation. *International Polymer Processing* 20, 162–169.
8. J. Zhao, R. H. Mayes, G. Chen, H. Xie, and P. S. Chan (2003) Effects of process parameters on the micromolding process. *Polymer Engineering and Science* 43, 1542–1554.
9. J. Zhao, R. H. Mayes, G. Chen, P. S. Chan, and Z. J. Xiong (2003) Polymer micromould design and micromoulding process. *Plastics, Rubber and Composites* 32, 240–247.
10. D. S. Kim, J. S. Kim, Y. B. Ko, J. D. Kim, K. H. Yoon, and C. J. Hwang (2008) Experimental characterization of transcription properties of microchannel

geometry fabricated by injection molding based on Taguchi method. *Microsystem Technologies* 14, 1581–1588.

11. K. Mönkkönen, J. Hietala, P. Päkkönen, E. J. Pääkkönen, T. Kaikuranta, T. T. Pakkanen, and T. Jääskeläinen (2002) Replication of submicron features using amorphous thermoplastics. *Polymer Engineering and Science* 42, 1600–1608.

12. S. Kang, J. Kim, and H. Kim (2000) Birefringence distribution in magneto-optical disk substrates fabricated by injection-compression molding. *Optical Engineering* 39, 689–694.

13. K. Seong, S. D. Moon, and S. Kang (2001) An optimum design of replication process to improve optical and geometrical properties in DVD–RAM substrates. *Journal of Information Storage and Processing Systems* 3, 169–176.

14. H. Schift, C. David, M. Gabriel, J. Gobrecht, L. J. Heyderman,W. Kaiser, S. Köppel, and L. Scandella (2000) Nanoreplication in polymers using hot embossing and injection molding. *Microlectronic Engineering* 53, 171–174.

15. Y. Kim, Y. Choi, and S. Kang (2005) Replication of high-density optical disc using injection mold with MEMS heater. *Microsystem Technologies* 11(7), 464–469.

16. M. Yoshii, H. Kuramoto, and K. Kato (1994) Experimental study of transcription of minute width grooves in injection molding. *Polymer Engineering and Science* 34, 1211–1217.

17. Y. Kim, J. Bae, H. Kim, and S. Kang (2004) Modeling of passive heating for replication of submicron patterns in optical disk substrates. *Journal of Physics D: Applied Physics* 37, 1319–1326.

18. B. H. Kim and N. P. Suh (1986) Low thermal inertia molding. *Polymer-Plastics Technology and Engineering* 25, 73–93.

19. M. J. Liou and N. P. Suh (1989) Reducing residual stresses in molded parts. *Polymer Engineering and Science* 29, 441–447.

20. B. Kim and M. F. Niemeyer (1995) *Insulated mold structure for injection molding of optical disks*, U.S. Patent No. 5,458,818.

21. K. Inoue, K. Hayashi, Y. Kawasaki, E. Ohno, S. Masuhara, and M. Kaneko (2003) Study on 100 Gbit/inch^2 density molding using double-sided heat insulated mold. *International Symposium on Optical Memory*, Japan.

22. Y. Kim, K. Seong, and S. Kang (2002) Effect of insulation layer on transcribability and birefringence distribution in optical disk substrate. *Optical Engineering* 41(9), 2276–2281.

23. H. Wang, B. Cao, C. K. Jen, K. T. Nguyen, and M. Viens (1997) Online ultrasonic monitoring of the injection molding process. *Polymer Engineering and Science* 37, 363–376.

24. T. Saito, I. Satoh, and Y. Kurosaki (2002) A new concept of active temperature control for an injection molding process using infrared radiation heating. *Polymer Engineering and Science* 42, 2418–2429.

25. CIMP (1994) *Consortium Meeting Notes*, Cornell University, November 3.

26. CIMP (1993) *Consortium Meeting Notes*, Cornell University, June 3.

27. D. Yao and B. Kim (2003) Developing rapid heating and cooling systems using pyrolytic graphite. *Applied Thermal Engineering* 23, 341–352.

28. S. V. Patanker (1980) *Numerical Heat Transfer and Fluid Flow*, Talyor & Francis, USA.

29. K. M. B. Jansen and J. van Dam (1992) An analytical solution for the temperature profiles during injection molding, including dissipation effects. *Rheologica Acta* 31, 592–602.

30. C. A. Hiebera and S. F. Shena (1980) A finite-element/finite-difference simulation of the injection molding filling process. *Journal of Non-Newtonian Fluid Mechanics* 7, 1–32.

31. K. M. B. Jansen and A. A. M. Flaman (1994) Construction of fast-response heating elements for injection molding applications. *Polymer Engineering and Science* 34, 894–897.

32. D. Yao and B. Kim (2002) Injection molding high aspect ratio microfeatures. *Journal of Injection molding Technology* 6, 11–17.

33. R. C. Luo and O. Chen (1998) MEMS-based thin-film pressure/temperature sensor for online monitoring injection molding, *Industrial Electronics Society, 1998. IECON '98. Proceedings of the 24th Annual Conference of the IEEE*, vol. 3, 1306–1309.

34. R. C. Luo, C. E. Lin, C. M. Chen, and Y. S. Chen (1999) Rapid resin mold with embedded thin-film pressure/temperature sensors. *Industrial Electronics Society, 1999. IECON '99 Proceedings. The 25th Annual Conference of the IEEE*, vol. 3, 1301–1306.

35. R. C. Luo and C. S. Tsai (2001) Thin-film PZT pressure/temperature sensory arrays for online monitoring of injection molding. *Industrial Electronics Society, 2001. IECON '01 Proceedings. The 27th Annual Conference of the IEEE*, vol. 1, 375–380.

36. S. Kang, C. A. Hieber, and K. K. Wang (1998) Optimum design of process conditions to minimize residual stress in injection-molded parts. *Journal of Thermal Stress* 21, 141–155.

37. K. K. Tan and J. C. Tang (2002) Learning-enhanced PI control of ram velocity in injection molding machines, *Engineering Applications of Artificial Intelligence* 15, 65–72.

38. F. Gao, Y. Yang, and C. Shao (2001) Robust iterative learning control with applications to injection molding process. *Chemical Engineering Science* 56, 7025–7034.

39. J. Choi, Y. Yamaguchi, S. Morales, R. Horowitz, Y. Zhao, and A. Majumdar (2003) Design and control of a thermal stabilizing system for a MEMS optomechanical uncooled infrared imaging camera. *Sensors and Actuators A: Physical* 104, 132–142.

40. Y. Kim, J. Kwon, Y. Choi, and S. Kang (2004) Replication of high-density optical disc using injection mold with MEMS heater. *Information Storage and Processing Systems Conference*, USA.

CHAPTER 5

Hot Embossing of Microstructured Surfaces and Thermal Nanoimprinting

5.1 INTRODUCTION

The increasing demand for micro-optical elements, in the fields of data storage, optical communication, functional surface, and digital display, has put mass fabrication technology for polymeric micro/nanopatterns at the forefront of research. Microlenses and microlens arrays can be produced using various methods such as the photoresist reflow method [1–3], etching [4,5], laser ablation [6,7], deposition [8,9] the microjet method [10,11], the photo thermal method [12], and microreplication [13–18]. And, nanopatterns are fabricated using electron beam lithography, focused ion beam and interferometer lithography methods explained in Chapter 2.

Among these, microreplication, including injection molding, compression molding, and hot embossing, is regarded as the most suitable mass-production process to replicate microlenses and nanopatterns because it offers high repeatability, mass producibility with low cost, and versatility in selecting polymers. Hot embossing developed by Becker and his colleagues uses a film-type material, which is placed on a hot plate in the embossing machine and heated above the glass transition temperature, then pressed to emboss the micropattern [14,15]. It was developed in the microcompression molding system with a silicon mold insert for a microlens with a diameter of 300 μm [16,17]. The microcompression molding system was modified from a conventional macrocompression molding system for micro-size replication. Here, various forms of polymers, such as powders, blocks, and films, can be used as replicated materials.

Micro/Nano Replication: Processes and Applications, First Edition. Shinill Kang.
© 2012 John Wiley & Sons, Inc. Published 2012 by John Wiley & Sons, Inc.

Among nano replication processes, nanoinjection molding, nanocompression molding, nanohot embossing, and nano-UV imprinting can be used to fabricate polymeric components that contain nanostructures, and these processes have been studied for several decades. It is noted that these processes do not require an additional etching process to form the final desired nanostructures. However, these processes have drawbacks when stack structures with fine alignment are required [18,19]. Another class of nano replication process is nanoimprint lithography, where the process has two basic steps of imprint and pattern transfer. In the imprint step, a nanomold is pressed into a resist on a substrate to form nanostructures, and in the subsequent pattern transfer step, usually reactive ion etching (RIE) is used to remove the residual resist in the compressed area [20–23]. Nanoimprint lithography has advantages when stack structures are necessary. Chou et al. [20,21] have applied a nanoimprint lithography process to produce nanoscale patterns for high-density data storage at low cost. Lebib et al. [22] have reported replication of nanopatterns over a four-inch wafer area using nanoimprint lithography. Hirai et al. [23] have proposed a solution for mass production using nanoimprint lithography with a metallic mold to solve problems such as the short lifetime of a silicon or quartz mold.

In micro/nano replication processes, micromold inserts, which contain the micro/nanopatterns, are required, and the quality of the mold inserts determines the success of whole process. A silicon mold insert is frequently used due to its ease of manufacture. However, a silicon mold insert is too brittle to be used for compression or injection molding for mass production, where high-pressure shock is applied to the mold cavity repeatedly. A metallic mold insert can provide a solution to this problem, and either mechanical machining or electroforming on any micropatterns can be used to make micropatterns on the metallic mold insert. However, it is difficult to make patterns of lens shape in a metallic mold insert by mechanical machining if the lens diameter is less than around 300 μm [24].

This chapter describes the thermal imprinting process used to fabricate microlenses and nanopatterns. The thermal imprinting process shown in Figure 5.1 is a fabrication process based on the reverse geometry of a mold with heat and pressure. Therefore, the interaction between the polymer and the mold is important. Optimizing the deformation of polymer patterns becomes more difficult as the size of the polymer decreases. This chapter discusses research to control the process conditions of thermal imprinting to improve the replication quality, and describes the fabrication of microlenses and nanopatterns using the micro/nanocompression molding system and processes that have been developed.

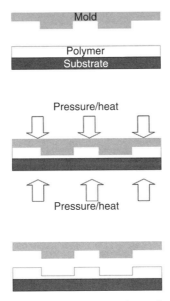

FIGURE 5.1   Fabrication procedure of nano replication technology with the nanomold for the polymeric nanopattern.

5.2 DEVELOPMENT OF MICROCOMPRESSION MOLDING PROCESS

Microcompression molding with powdered optical polymer was used to fabricate a polymeric micro/nanopatterns. A micro/nanocompression molding system was specially designed for this study as shown in Figure 5.2. The outer cylinder has electrical heating elements for mold heating. A thermocouple and a load cell were placed in the upper ram to measure the mold temperature and pressure histories for the control of temperature and pressure. The mold insert jig was placed on the lower ram and the mold insert was mounted on the jig. The jig was made to have a highly flat mirror surface to prevent uneven pressure distribution that can cause cracks in the silicon mold insert. Microcompression molding process is similar to the general replication process. The micro/nanocompression molding process progressed in four stages as follows:

(1) Replication material preparation. An optical grade PMMA powder, with a transparency of 93.0% at 3.2 mm thickness, a refractive index of 1.489 at 655 nm wavelengths, a haze of 1.0%, and a glass transition temperature of 110°C, was placed on the mold insert.

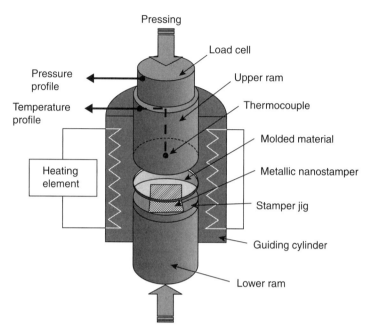

FIGURE 5.2   Microcompression molding system. Reprinted with permission from Ref. [25]. Copyright 2003, Institute of Physics.

(2) Mold heating and pressing. The compression pressure and replication temperature are governing factors. The mold was heated to a replication temperature that is above the glass transition temperature of PMMA. During heating to replication temperature, prepressure was applied to maintain contact between the melting powder and the mold insert. The prepressure helps powder to fill the cavity and means that it is heated effectively. Since the powder particles are diffusion-bonded while filling the micropatterns in microcompression molding, the replication temperature must be raised to above the glass transition temperature of the polymer. When the temperature of the mold reached the replication temperature, the compression pressure was applied. The replication pressure had to be high enough to improve the replication quality and to give the required bonding force between adjacent polymer particles. However, if the replication temperature and the compression pressure are raised excessively, Not only can it deteriorate the mechanical and optical qualities of the replicated parts but it can also cause various defects.

(3) Cooling and packing. The mold was cooled while the compression pressure was maintained. A proper cooling rate is important to ensure the quality of the replicated lens.

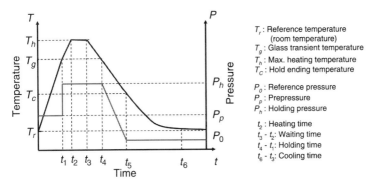

FIGURE 5.3   Relationship between the temperature and compression force during micro/nanothermal imprinting process.

(4) Releasing. Once the demolding temperature had been reached, the replicated microlenses were released from the mold insert.

The graph in Figure 5.3 shows the relationship of temperature and pressure in the micro/nanothermal imprinting process previously described.

5.3   TEMPERATURE DEPENDENCE OF ANTI-ADHESION BETWEEN A MOLD AND THE POLYMER IN THERMAL IMPRINTING PROCESSES

Many technical problems in nanomolding processes have to be overcome to mass produce components at low cost. For example, as the features of the nanopatterns on the mold become smaller, one needs to raise the mold temperature. The reason is that above the glass transition temperature the polymer has a higher fluidity than that below and, therefore, can better fill the submicron patterns. However, when the mold temperature increases excessively, the polymer may stick to the mold, worsening the replication quality as well as the surface quality of molded components with nanopatterns, as described in detail by Kang and co-workers [21,26]. Therefore, it is necessary to analyze quantitatively the effects of the mold temperature on the anti-adhesion property between the mold and the polymer in a pliable state, especially at temperatures above the glass transition temperature of the polymer.

The anti-adhesion property of the mold can be evaluated using contact angle measurement using diagnostic liquids at room temperature [27–30]. However, diagnostic liquids cannot be used at high temperatures since they evaporate. Another drawback to the method using diagnostic liquids is that it cannot analyze the anti-adhesion property between the mold and the polymer precisely. An alternative method is to use the polymer in a pliable state above the glass transition temperature. Only a few researchers have measured the contact angle by using the polymer in a pliable state because the high viscosity and the thermal instability of the polymer in a pliable state have made experiments difficult. Grundke et al. [31] have developed a technique for measuring the wetting tension of the polymer in the temperature range between the glass transition temperature and the melting temperature using the Wilhelmy balance method. Wulf et al. [32] have analyzed the surface tension and the density of polymer melts as a function of the temperature below the melting temperature by using the sessile drop method. However, the anti-adhesion property between the actual mold and the polymer has not been analyzed. In the analysis of the anti-adhesion property between the mold and the polymer, the contact angle should be evaluated at the actual replication temperature. This study presents an experimental method of analyzing the anti-adhesion property between a mold with submicron patterns and a polymer in a pliable state. For analysis of the anti-adhesion property between the mold and the polymer, the contact angle was measured at the actual replication temperature. In this process, the mold was heated above the glass transition temperature of the polymer. Finally, as a practical example, the correlation between the contact angle and the surface quality of the replicated substrates using a mold with submicron patterns as a function of replication temperature was obtained quantitatively.

5.3.1 Defects in Replicated Micro-Optical Elements

The dimensional accuracy and the surface quality become more critical in micro-optical elements than in conventional replicated parts. The quality of microreplicated optical elements was sensitive to replication process conditions. Figure 5.4 shows defects, which occur usually during microcompression molding process. Figure 5.4a shows the defect due to slip by microvibration of the mold and the nonparallel release, which could have been avoided by proper design of the jig and the ejecting units. The burst and the tear shown in Figure 5.4b developed when the excessive compression pressure was applied to the polymer in the filling stage and cooling stage. On the other hand, Figure 5.4c shows the defect due to shrinkage by insufficient pressure in the cooling stage. These defects could be eliminated by redesigning the mold and modifying the processing conditions.

5.3.2 Analysis of Polymer in Process Condition of Thermal Imprinting

Figure 5.5 shows a schematic diagram of the contact angle measurement system. This system analyses the anti-adhesion property between a stamper

(a)

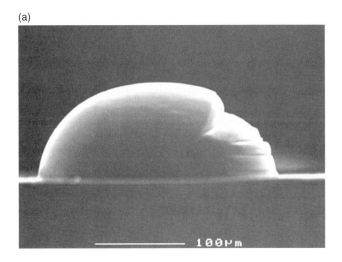

(b)

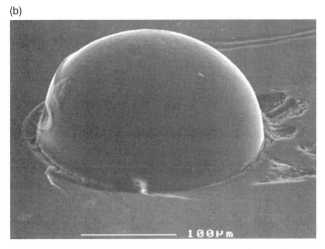

FIGURE 5.4 Various defects of replicated micro lens: (a) slip; (b) burst; (c) shrinkage. Reprinted with permission from Ref. [33]. Copyright 2004, APEX/JJAP.

with submicron patterns and a polymer in a pliable state. This system consists of a hot plate and a thermal chamber, which are used to heat the air in the thermal chamber up to the temperature of the actual replication process. The hot plate was connected to the temperature controller, and the temperature of the mold surface in the thermal chamber was measured using a thermometer. A microscope and a CCD camera were used to record the images of the polymer in a pliable state on the specimen of the mold during the measurement, and these images were analyzed using an image extract or to extract the contact angle (ImagePRO300.SEO.Co).

(c)

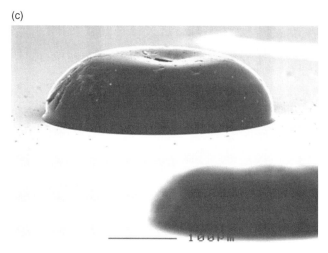

FIGURE 5.4 (*Continued*)

To evaluate the contact angle of the mold and measure the surface quality and replication quality of the replicated substrates at various replication temperatures, a mold with submicron pillar patterns (pitch = 500 nm) was used. The mold with submicron pillar patterns was fabricated as follows. As a first step, the master, with a submicron hole pattern, was fabricated using electron beam (*e*-beam) lithography. After the Ni seed layer was deposited on the master patterns using *e*-beam evaporation, the metallic mold was fabricated using a Ni-electroforming process.

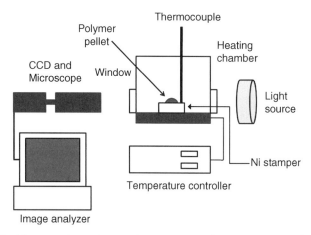

FIGURE 5.5 The experimental setup for contact angle measurements. Reprinted with permission from Ref. [34]. Copyright 2004, Institute of Physics.

(a)

(b)

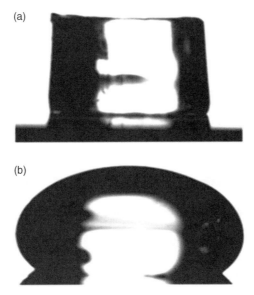

FIGURE 5.6 CCD image of polymer pellet: (a) before softening; (b) after softening to a pliable state. Reprinted with permission from Ref. [34]. Copyright 2004, Institute of Physics.

Polymethylmethacrylate (PMMA) was used as the polymer material, and it was used in the measurement of the contact angle on the mold. The contact angle was measured using the polymer in a pliable state at a temperature above the glass transition temperature, Tg, of the polymer. With the Tg of polymer material known, the minimum temperature that can be used in measuring the contact angle of the mold was determined at around 180°C, experimentally. When the temperature of the hot plate in the contact angle measurement system is maintained at this temperature for about 30 min, the polymer pellet transforms to a hemisphere. Figure 5.6 shows CCD images of a PMMA pellet (a) before softening and (b) after softening to a pliable state, respectively. Additionally, considering the mold temperature in the actual replication process, the peak mold temperatures were set to 200, 210, 220, 230, 240, and 250°C to observe the effect of temperature on the wettability of the mold surface and the effect of wettability on the surface quality of replicated substrates. The thermal chamber was first heated to 180°C and maintained for 30 min, which is the initial temperature for contact angle measurement. At this temperature, the thermal chamber was further heated to the peak mold temperature and maintained at the peak mold temperature. The chamber was heated at 10°C min^{-1} during the heating stage, and the isothermal stage lasted 15 min. After the heating and isothermal stages, the specimen of the mold in the thermal chamber was cooled. During each experiment, the contact angle and the temperature of the mold were measured simultaneously every minute.

Figure 5.7 shows the histories of the mold temperature and the contact angle between the mold and the PMMA during each experiment for different peak mold temperatures. As shown in Figure 5.7, during the heating and

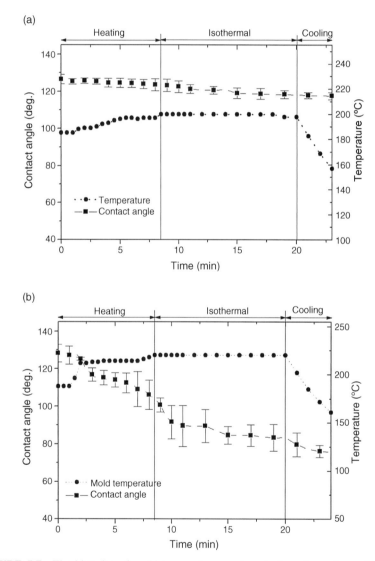

FIGURE 5.7   The histories of mold temperatures and contact angle on the mold of the PMMA for different peak. Reprinted with permission from Ref. [34]. Copyright 2004, Institute of Physics.

(c)

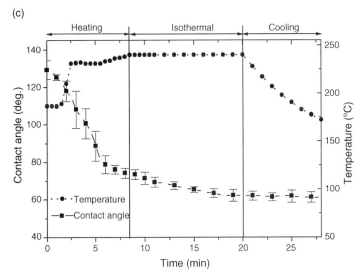

FIGURE 5.7 (*Continued*)

the isothermal stages, the contact angle between the mold and the polymer in a pliable state decreased continuously, but during the cooling stage, the contact angle remained constant. This result shows that the terminal contact angles after the cooling stage depend on the peak mold temperature. In other words, this behavior indicates that the wettability of the mold in the demolding process depends on the molding temperature and not on the demolding temperature.

5.3.3 Analysis of Replication Quality Fabricated in Different Peak Temperature

The anti-adhesion property of the mold could be predicted by measuring the contact angle between the mold and the polymer at the temperature of the actual replication process. However, the anti-adhesion property and replication quality of the mold with submicron patterns could not be described directly by measuring the contact angle. So, as a practical example, the surface roughness and the replication quality of the replicated substrates were evaluated by using nanocompression molding at different replication temperatures. The replication temperatures were the same as the peak mold temperature in the contact angle measurement in the range 180–250°C. Nanocompression molding with powdered PMMA was used to fabricate a plastic substrates. After powdered PMMA was put on the mold, the mold was heated to the replication temperature. During heating, a small prepressure

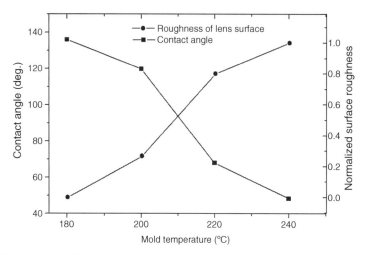

FIGURE 5.8   The effect of mold temperature on the surface roughness of the replicated substrate and contact angle of the mold. Reprinted with permission from Ref. [34]. Copyright 2004, Institute of Physics.

was applied to maintain contact between the melted powder and the mold. Compression was applied to replicate the polymeric substrates when the mold temperature reached the replication temperature. The mold was cooled, while maintaining the compression pressure.

The surface roughness of the replicated substrate was measured using a surface profiler (DEKTAK6M). Figure 5.8 shows the effect of mold temperature on the surface roughness of the replicated substrate and the contact angle of the mold. The surface roughness (rms) of the replicated substrates was 2.8–3.5 nm. As shown in Figure 5.8, as the mold temperature increased, the surface roughness of the replicated substrates increased, while the contact angle value of the mold decreased. The surface roughness tended to be inversely related to the contact angle value. In other words, as the mold temperature increased, the surface roughness tended to rise, corresponding to the falling trend of the contact angle. This result shows the fact that an increase in the wettability of the mold surface worsens the anti-adhesion property between the mold and the polymer. This deterioration of the anti-adhesion property between the two surfaces again results in deterioration of the surface quality of the replicated substrates. Additionally, as shown in Figure 5.8, when the mold temperature is around the T_m of the PMMA, 200–220°C, the contact angle value decreases and the surface roughness increases markedly. The increase in the wettability of the mold surface around T_m seriously worsened the surface quality of the replicated substrates. Figure 5.9 shows the microlens shapes fabricated under different thermal

(a)

(b)

FIGURE 5.9 Comparison of surface quality of replicated microlenses: (a) 160°C (b) 200°C (c) 240°C. Reprinted with permission from Ref. [25]. Copyright 2003, Institute of Physics.

conditions. The microlens was not replicated perfectly at a replication temperature of 106°C due to underfilling. In addition, the surface quality of the replicated lens deteriorated due to sticking, as shown in Figure 5.9c. At the proper replication temperature of 200°C, the patterns were successfully reproduced, as shown in Figure 5.9c.

Atomic force microscopy (AFM) was used to measure the submicron patterns in the metallic mold and polymeric substrates replicated at 180, 200,

(c)

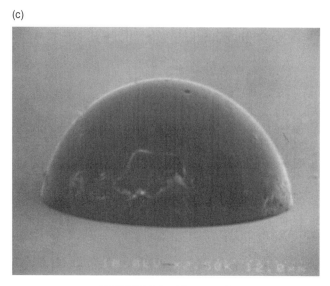

FIGURE 5.9   (*Continued*)

220, and 240°C. The area of the actual scanning by AFM was $5 \times 5\,\mu m^2$ for each measurement point. Figure 5.10 shows the profile obtained from AFM data of the mold with submicron pillar pattern. Figure 5.11 shows the profiles obtained from AFM data of the replicated patterns at 180, 200, 220, and 240°C. As shown in Figures 5.11a and b, below 200°C, the polymer could not fill the submicron pattern in the mold completely. And as shown in Figure 5.11c, polymeric components were replicated with good surface quality around 220°C. As shown in Figures 5.11b and c, when the mold temperature was around the Tm of the PMMA, 200–220°C, the replication quality was improved markedly since the elevation of the mold temperature increased the fluidity of the polymer. However, as shown in Figure 5.11d, when the replication temperature is above 240°C, the

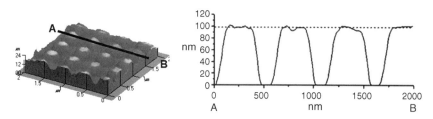

FIGURE 5.10   AFM image of a metallic mold with submicron pillar patterns (100 nm height of the submicron pillar pattern in the mold). Reprinted with permission from Ref. [34]. Copyright 2004, Institute of Physics.

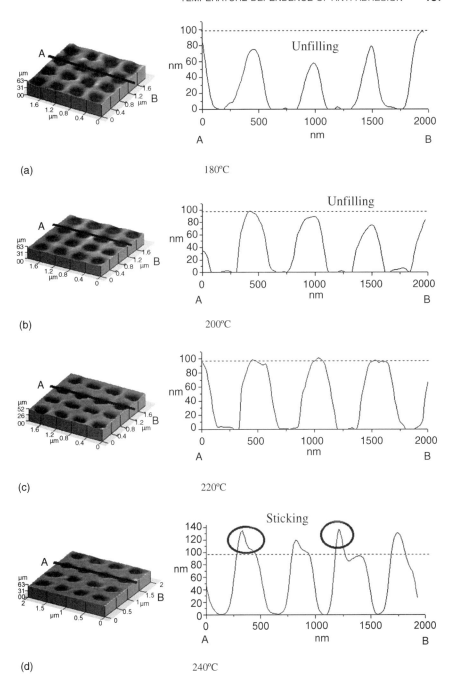

FIGURE 5.11   AFM images of a polymer pattern with submicron hole patterns for different mold temperatures: (a) 180°C; (b) 200°C; (c) 220°C; (d) 240°C. Reprinted with permission from Ref. [34]. Copyright 2004, Institute of Physics.

sticking problem between the metallic mold and the polymer worsens the surface profile and surface quality of nanohole patterns in replicated substrates.

5.4 FABRICATION OF A MICRO-OPTICS USING MICROCOMPRESSION MOLDING WITH A SILICON MOLD INSERT

5.4.1 Fabrication of Microlens Components Using Si Mold Insert

The increasing demand for microlenses in the fields of optical data storage, optical communication, and digital displays has made the development of fabrication technology for polymeric microlenses a priority. Many different methods other than conventional methods such as turning and polishing have been proposed to produce individual microlenses or microlens arrays. In this section, thermal imprinting is described using the thorough Si mold etching process introduced in Chapter 2.

Etched silicon is used directly as the mold insert instead of metallic materials to produce hemispherical microlenses and four-step diffractive optical elements. Various defects during this process were analyzed, and the results were used to control the main process parameters of mold temperature and compression pressure to optimize the process experimentally. Refractive microlenses with a 153 μm radius of curvature were fabricated. Blazed and four-step diffractive optical elements with a 24 μm pitch and 5 μm depth were also fabricated.

The various defects can be prevented by properly managing the replication process and carefully preparing the mold insert and mold base. The temporal histories of the mold temperature and compression pressure were determined experimentally to be the dominant governing process conditions. An increase in fluidity, which is affected by the viscosity of the molten polymer, can improve the replication quality; this can be achieved by increasing the replication temperature. Furthermore, because the powder particles are diffusion-bonded while filling the micropatterns in microcompression molding using polymer powders, the replication temperature must be greater than the glass transition temperature of the polymer. The compression pressure must also be high enough to improve the replication quality and provide the required bonding force between adjacent polymer particles. However, if the replication temperature and the compression pressure are too high, they degrade the optical quality of the replicated micro-optics and cause various defects, as explained earlier. Figures 5.12 and 5.13 show scanning electron

(a)

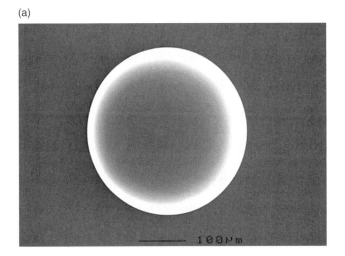

(b)

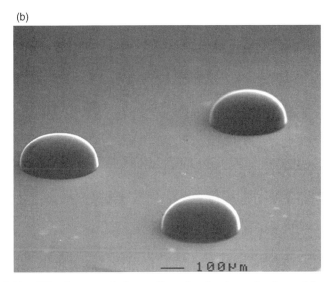

FIGURE 5.12 SEM images of the replicated refractive microlens: (a) a microlens; (b) microlenses. Reprinted with permission from Ref. [33]. Copyright 2004, APEX/JJAP.

microscope (SEM) images of the micro-optical elements replicated with proper control of the processing conditions.

5.4.2 Analysis of Refractive Microlens

The surface profile of the mold insert and the replicated microlens were measured interferometrically using an interferometric three-dimensional profiler. The surface quality and the radius of curvature of only the top portion of

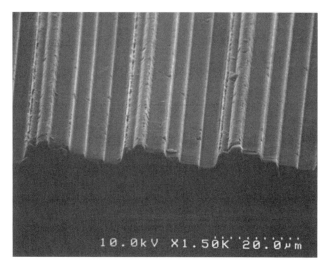

FIGURE 5.13   SEM images of replicated diffractive optical elements. Reprinted with permission from Ref. [33]. Copyright 2004, APEX/JJAP.

the lens surface where the light path passes were scanned due to the limitations of the profiler. Figure 5.14 shows three-dimensional surface profiles of the mold insert and replicated lens. The radius of curvature of the hemispherical pit on the mold insert was 153 μm. The maximum deviation of the radius of curvature was within ±5 μm. The radius of curvature of the replicated microlens was 153 μm, the same as the mold insert. Figure 5.15 shows the ideal hemispherical profile with a radius of 153 μm, the profile of the mold insert, and the profile of the replicated microlens. This shows the symmetric lens profile with a smooth surface, and illustrates that the replicated lens replicates the hemispherical pattern closely. The refractive index of the replicated microlens, measured using a Metricon 2010 Prism Coupler, was 1.4914. An atomic force microscopy calibration specimen of 10 μm square patterns was used to characterize the optical properties of the replicated microlenses. The microlens was capable of magnifying an image by 3.143 times.

5.5   FABRICATION OF A MICROLENS ARRAY USING MICROCOMPRESSION MOLDING WITH AN ELECTROFORMING MOLD INSERT

5.5.1   Fabrication of Microlens Components Using Ni Mold Insert

Micromold inserts, which contain the micropatterns, are required in microreplication processes. The quality of the mold inserts determines the

(a)

(b)

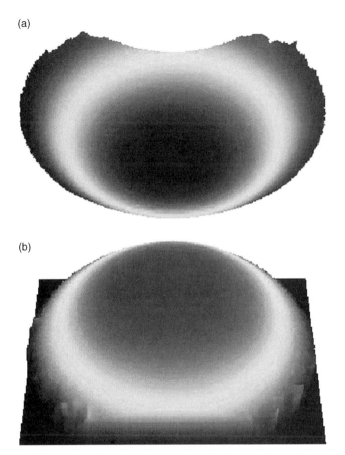

FIGURE 5.14   Three-dimensional surface profiles of: (a) mold insert; (b) replicated lens, measured with three-dimensional profiler. Reprinted with permission from Ref. [17]. Copyright 2002, SPIE.

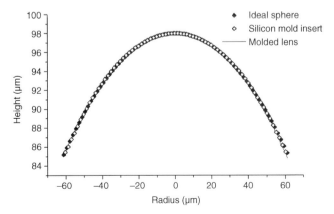

FIGURE 5.15   Comparison among ideal sphere profile, silicon mold insert profile, and replicated lens profile. Reprinted with permission from Ref. [17]. Copyright 2002, SPIE.

(a)

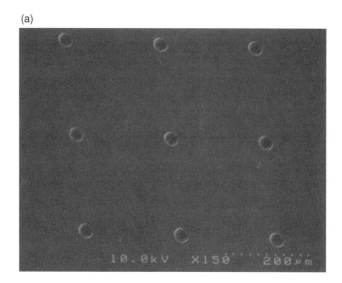

(b)

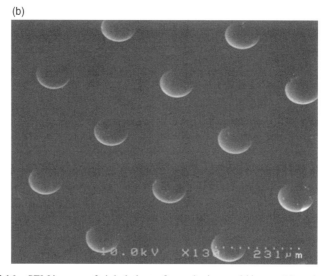

FIGURE 5.16   SEM images of nickel electroformed micromold insert: (a) cavity diameter of 36 μm; (b) cavity diameter of 96 μm. Reprinted with permission from Ref. [25]. Copyright 2003, Institute of Physics.

success of the entire process. A silicon micromold insert is frequently used because it is easy to manufacture. However, a silicon mold insert is too brittle to be used for compression or injection molding for mass production where high-pressure shock is repeatedly applied to the mold cavity. A metallic mold insert can provide a solution to this problem, and either mechanical machining or electroforming on any micropattern can be used to create micropatterns on the metallic mold insert. However, making patterns of the lens shape in a metallic mold insert by mechanical machining is difficult if the lens diameter is less than 300 µm.

The main objective of this work is to design and fabricate a microlens array using microcompression molding with a metallic mold insert. Microlenses with diameters of 36–96 µm, radii of curvature of 20–60 µm, and a pitch of 250 µm have been fabricated. The reflow method and an electroforming process were used to create the mother and the metallic mold insert, respectively, because mechanical machining could not be used. Microcompression molding with powder polymer was used to replicate the microlenses. Finally, the surface profiles, imaging qualities, and surface roughness of the replicated lenses were measured and analyzed.

Figure 5.16 shows SEM images of electroformed nickel mold inserts. The diameters of the concave surfaces were 36 and 96 µm, and the pitch of the array was 250 µm in both cases. Microcompression molding with powdered optical polymer was used to fabricate a polymeric microlens array. Figure 5.17 shows SEM images of the replicated microlens and lens array with diameters of 36 and 96 µm, respectively.

5.5.2 Analysis of Replication Quality

The surface profiles of the mother and replicated lenses were measured using a three-dimensional optical profiler, a mechanical profiler, and an SEM. The optical interferometric profiler could measure the profile at the top portion of the lens surface with high resolution, but the outer portion of the lens surface could not be measured precisely due to a high tangential angle, whereas the mechanical profiler could not find the exact center profile of the lens. Therefore, the lens profiles were generated by superposition of mechanical and optical measurements. Table 5.1 shows the diameter, the radius of curvature at the lens center, and the sag height H of the mother and the replicated lenses. The radius of curvature and the sag height of the replicated lenses deviated from those of the mother lenses by less than 0.8 µm (1.3%) and 0.2 µm (0.6%), respectively. Figure 5.18 shows the comparison between surface profiles of the mother lenses and those of replicated lenses with diameters of 36 and 96 µm, respectively.

The image spot intensity distribution was measured by a beam profiler using a 665 nm laser source. The sample microlens was the replicated lens with a diameter of 96 µm and a radius of curvature of 58.29 µm. The intensity profile at focal point of the replicated microlens is shown in Figure 5.19. The spot diameter with $1/e^2$ intensity of the peak was 1.386 µm. The focal length was measured as 125 µm for the replicated lens with a diameter of 96 µm and a radius of curvature of 58.29 µm.

(a)

(b)

FIGURE 5.17   SEM images of replicated microlens array: (a), (c) lens diameter of 36 μm; (b), (d) lens diameter of 96 μm. Reprinted with permission from Ref. [25]. Copyright 2003, Institute of Physics.

The surface quality of the mold insert defines the final surface quality of the replicated lens. An atomic force microscope (AFM) was used to measure the surface roughness of the mold insert and replicated lenses. The specimens were randomly selected to avoid the possibility of a systematic error infiltrating the system. The actual scanned area for each measurement point on the curved surfaces was 5×5 μm^2. Figure 5.20 shows AFM images of the cavity surface and the replicated surface. The surface roughness (RMS) of

(c)

(d)

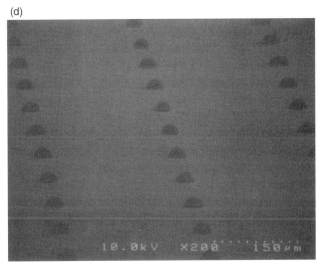

FIGURE 5.17   (*Continued*)

the metal mold cavity was 0.95 nm, which guarantees a mirror-surface mold cavity. The surface roughness (RMS) of the replicated lens was 3.98 nm. These measurements show the possibility of using the present replicated lens for applications such as data storage and optical communication, in which the wavelengths between 405 and 850 nm are used. It is because RMS of 3.98 nm for the present replicated lens is about 0.01λ for high-density optical data storage applications ($\lambda = 405$ nm) and about 0.005λ for optical communications ($\lambda = 805$ nm).

TABLE 5.1 Comparison of the Radius of Curvature and the Sag Height of the Mother and Molded Lenses Diameter of Lens

	Mother Lens		Replicated Lens			
Diameter of Lens $(D, \mu m)$	Radius of Curvature at Lens Center $(R_c, \mu m)$	Sag Height at Center $(H, \mu m)$	Radius of Curvature at Lens Center $(R_m, \mu m)$	Sag Height at Center $(H_m, \mu m)$	E_1 (%)	$E_2(\%) = \lvert (H - H_m)/H \rvert \times 100$
96	59.06	25.24	58.29	25.08	1.30	0.6
76	44.53	23.46	43.44	23.46	0.31	0
56	31.38	21.61	31.02	21.61	1.15	0
46	24.90	19.85	25.04	19.93	0.56	0.4
36	19.95	17.49	19.75	17.45	1.00	0.2

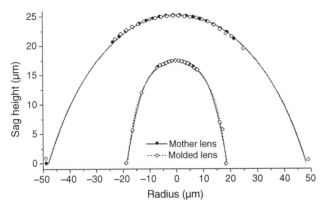

FIGURE 5.18 Comparison between the mother and replicated lens profiles. Reprinted with permission from Ref. [25]. Copyright 2003, Institute of Physics.

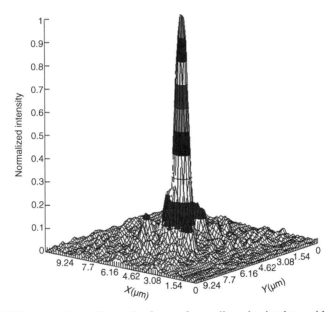

FIGURE 5.19 Intensity profile at the focus of a replicated microlens with a diameter of 96 μm. Reprinted with permission from Ref. [25]. Copyright 2003, Institute of Physics.

5.6 APPLICATION OF MICROCOMPRESSION MOLDING PROCESS

5.6.1 Fabrication of a Microlens Array Using Microcompression Molding

As an example of this micro/nanothermal forming, microcompensatory lens was designed and fabricated for the small form factor optical data

(a)

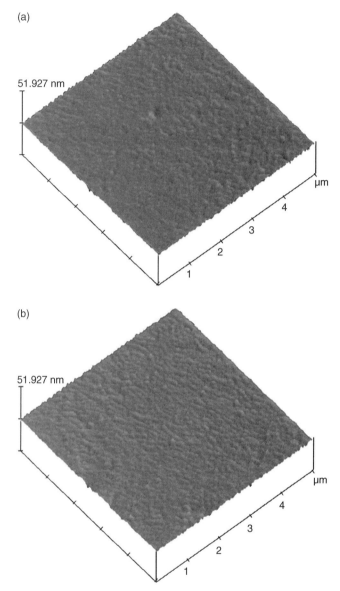

(b)

FIGURE 5.20   AFM images of (a) nickel mold insert and (b) replicated lens. Reprinted with permission from Ref. [25]. Copyright 2003, Institute of Physics.

storage [35]. The designed optical path for small form factor optical pickup is compatible with the Blu-ray disk format in terms of the wavelength of laser diode and numerical aperture, and the thickness of the cover layer. The schematic diagram of the optical path is shown in Figure 5.21. The optical path and its optical elements are designed using commercial optical design software.

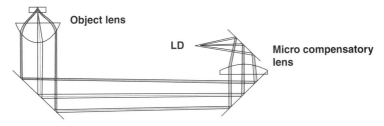

FIGURE 5.21 Schematic diagram of the optical path. Reprinted with permission from Ref [33]. Copyright 2004, APEX/JJAP.

It is difficult for single objective lens to compensate the various aberrations; therefore a microcompensatory lens shown in Figure 5.22 was designed in order to complement the objective lens in compensating various aberrations. It was used to improve the optical performance by disrupting the light emission and changing the optical overall length. Figure 5.23 shows the electroformed mold insert and replicated microcompensatory lens by micro/ nanocompression molding. The deviation of the surface profile of the replicated microcompensatory lens from designed profile is shown in Figure 5.24. The shape deviation was less than about 0.6 mm. The focal length and wavefront aberration of the optical system using this microcompensatory lens were designed as 1.4462 mm and 0.0200λ ($\lambda = 405$ nm), respectively. The measured focal length and wavefront aberration were 1.4462 mm and $0:0202\lambda$ ($\lambda = 405$ nm), respectively.

5.6.2   Fabrication of Metallic Nanomold and Replication of Nanopatterned Substrate for Patterned Media

In this section, another approach is described to apply the nano replication technology to patterned media without an RIE process. In particular, without an RIE process, nanopatterns of holes and pillars onto the polymeric substrate were successfully transferred using the Ni nanostamper by the precise control of the nano replication process. The replicated patterns were as small as 100 nm in diameter and 100 nm in depth with a surface roughness of less than 1 nm.

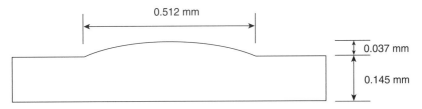

FIGURE 5.22 Design of microcompensatory lens. Reprinted with permission from Ref. [33]. Copyright 2004, APEX/JJAP.

(a)

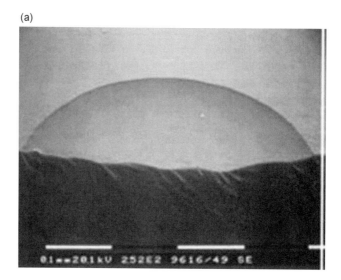

(b)

FIGURE 5.23   SEM images of (a) fabricated mold insert; (b) replicated microcompensatory lens. Reprinted with permission from Ref. [33]. Copyright 2004, APEX/JJAP.

A metallic nanostamper was fabricated using electroforming, and then the nanopatterned substrates were replicated using a nano replication process with the metallic nanostamper. After electroforming, the silicon wafer and *e*-beam resist were removed. The nanoscale pillars were successfully transferred from the holes of master nanopatterns. Figure 5.25 shows SEM and AFM images of the metallic nanostamper with a diameter of 200 nm, a pitch of 500 nm, and a depth of 100 nm.

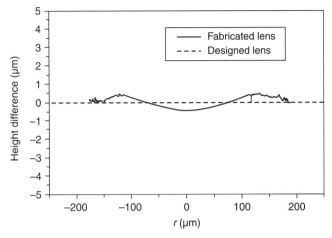

FIGURE 5.24 Height difference profile of fabricated microcompensatory lens from designed profile. Reprinted with permission from Ref. [33]. Copyright 2004, APEX/JJAP.

The metallic nanomold was used for a replication process. The replication process has been regarded as a suitable mass-production process for replicating nanopatterns. In a replication process, replication temperature and compressive pressure are the most dominant governing process parameters that determine the replication quality of replicated parts. If replication temperature and compressive pressure are too low, replicated patterns cannot be made because PMMA powders are not bonded and polymeric nanopatterns do not fill a metallic nanomold. In contrast, if replication temperature and compressive pressure are excessively high, they can deteriorate the replication quality because of the sticking effect. Therefore, it is important to properly control replication temperature and compressive pressure to improve the quality of replication in polymeric nanopatterns. It was found that replication temperature affects the quality of replication more than

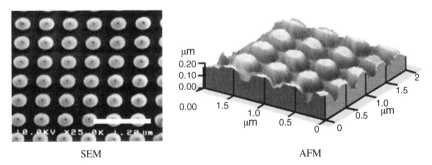

SEM   AFM

FIGURE 5.25 SEM and AFM images of the metallic nanomold with a diameter of 200 nm, a pitch of 500 nm and a depth of 100 nm. Reprinted with permission from Ref. [36]. Copyright 2004, Institute of Physics.

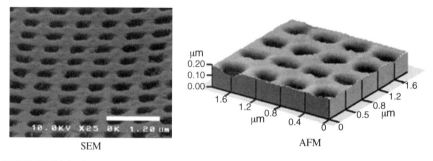

FIGURE 5.26   SEM and AFM images of the polymeric hole patterns with a diameter of 200 nm, a pitch of 500 nm, and a depth of 100 nm. Reprinted with permission from Ref. [36]. Copyright 2004, Institute of Physics.

compressive pressure does. The replication process was carried out at a temperature of around 210–220°C and a compressive pressure in the range of 2–4 MPa. To evaluate the quality of replication in the replicated patterns, the surface profiles of the polymeric hole patterns were analyzed by SEM and AFM as shown in Figure 5.26. Figure 5.26 shows SEM and AFM images of the polymeric hole patterns with a diameter of 200 nm, a pitch of 500 nm and a depth of 100 nm. The surface roughness of the hole patterns was ~8 Å. The typical surface roughness of the master nanopatterns, the metallic nanomold and the polymeric nanopatterns is summarized in Table 5.2. The SEM and AFM images confirm that the replicated parts contained the reverse polymeric patterns of the metallic nanomold. Further improvement of the surface quality of the patterns seems to be possible. For example, an anti-adhesion layer such as a self-assembled monolayer (SAM), described in Chapter 3, can reduce the sticking problem between mold and patterns, which leads to the enhancement of the surface quality of the patterns. As a matter of fact, this study showed that applying the anti-adhesion SAM to the nickel mold has significantly improved the surface roughness of the patterns.

To apply our method to patterned media, the nanopatterned substrate with pillar patterns was replicated. The nanopatterned substrate had pillar patterns with a diameter of 200 nm and a pitch of 500 nm, and a diameter of 100 nm and a pitch of 250 nm, respectively. Figures 5.27 and 5.28 show SEM and

TABLE 5.2   Surface Roughness of the Master Nanopatterns, the Metallic Nanopatterns and the Polymeric Nanopatterns

	Ra (Å)
Master nanopatterns	3.29
Metallic nanomolds	6.42
Polymeric nanopatterns	7.55

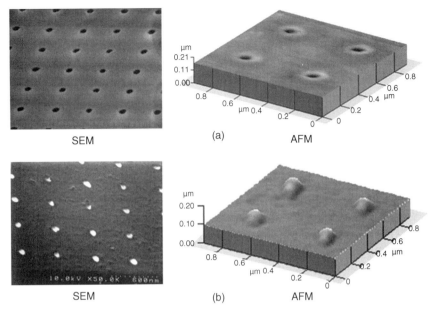

FIGURE 5.27 SEM and AFM images of (a) the metallic nanomold; (b) the nanopatterned substrate with a diameter of 200 nm and a pitch of 500 nm for patterned media. Reprinted with permission from Ref. [36]. Copyright 2004, Institute of Physics.

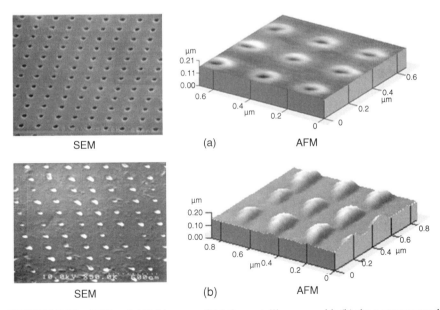

FIGURE 5.28 SEM and AFM images of (a) the metallic nanomold; (b) the nanopatterned substrate with a diameter of 100 nm and a pitch of 250 nm for patterned media. Reprinted with permission from Ref. [36]. Copyright 2004, Institute of Physics.

AFM images of the metallic nanostamper and nanopatterned substrate for patterned media.

Comparing patterns on the replicated polymeric substrate, we can find that the pillar patterns were not stable. In particular, when the pillar diameter was 100 nm, some defects were found on the substrate, and the patterns were not uniformly distributed as designed. From those results, it is expected that the replication of pillar patterns is more difficult than that of hole patterns probably due to nonfilling and sticking problems. It is necessary in the nano replication process that the process conditions should be further optimized and the releasing method should be further enhanced to solve those problems.

REFERENCES

1. Z. D. Popovic, R. A. Sprague, and G. A. N. Connell (1988) Technique for monolithic fabrication of microlens arrays. *Applied Optics* 27, 1281–1284.

2. Y. Lin, C. Pan, K. Lin, S. Chen, J. Yang, and J. Yang (2001) Polyimide as the pedestal of batch fabricated microball lens and micromushroom array. *Micro Electro Mechanical Systems, 14th IEEE International Conference*, 337–340.

3. S. Sinziger and J. Jahans (1999) *Micro-optics*, Wiley, New York, 85–93.

4. M. B. Stern and T. R. Jay (1994) Dry etching for coherent refractive microlens array. *Optical Engineering* 33, 3547–3551.

5. F. W. Ostemayer, Jr., P. A. Kohl, and R. H. Burton (1983) Photoelectrochemical etching of integral lenses on InGaP/InP. *Applied Physics Letters*, 43, 642–644.

6. M. Wakaki, Y. Komachi, and G. Kanai (1998) Microlenses and microlens arrays formed on a glass plate by use of a CO_2 laser. *Applied Optics* 37, 627–631.

7. S. Mihailov and S. Lazare (1993) Fabrication of refractive microlens array by eximer laser ablation of amorphous Teflon. *Applied Optics* 32, 6211–6218.

8. M. Kubo and M. Hanabusa (1990) Fabrication of microlenses by laser chemical vapor deposition. *Applied Optics* 29, 2755–2759.

9. Y. Fu and B. K. A. Ngoi (2001) Investigation of diffractive-refractive microlens array fabricated by focused ion beam technology. *Optical Engineering* 40, 511–516.

10. D. L. MacFarlane, V. Narayan, W. R. Cox, T. Chen, and D. J. Hayes (1994) Microjet fabrication of microlens array. *IEEE Photon. Technology Letters* 6, 1112–1114.

11. D. J. Hayes and W. R. Cox (1998) Microjet printing of polymers for electronics manufacturing. *Adhesive Joining and coating Technology in Electronics Manufacturing*. 168–173.

12. N. F. Borrelli, D. L. Morse, R. H. Bellman, and W. L. Morgan (1985) Photolytic technique for producing microlenses in photosensitive glass. *Applied Optics* 24, 2520–2525.

13. K. Seong, S. Moon, and S. Kang (2001) An optimum design of replication process to improve optical and geometrical properties in DVD-RAM substrates. *Micro and Nanosystems and Information Storage and Processing Systems* 3, 169–176.

14. P. Ruther, B. Gerlach, J. Gottert, M. Ilie, J. Mohr, A. Muller, and C. Obmann (1997) Fabrication and characterization of microlenses realized by a modified LIGA process. *Pure Applied Optics* 6, 643–653.

15. N. Moldovan, M. Ilie, N. Dumbravescu, M. Danila, A. Vitriuc, P. Sindile, and J. Mohr (1997) LIGA and alternative techniques for micro-optical components. *CAS '97 Proceedings IEEE*, 149–152.

16. S. Moon, S. Ahn, S. Kang, D. Choi, and T. Je (2001) Fabrication of refractive and diffractive plastic micro-optical components using microcompression molding. *Device and Process Technologies for Microelectronics, MEMS II*, 4592, 140–147.

17. S. Moon, S. Kang, and J. Bu (2002) Fabrication of polymeric microlens of hemispherical shape using micromolding. *Optical Engineering* 41, 2267–2270.

18. Y. Kim, N. Lee, Y.-J. Kim, and S. Kang (2004) Fabrication of metallic nanomold and replication of nanopatterned substrate for patterned media. *The Nanotechnology Conference 2004* (Boston) 3, 452–455.

19. S. Choi, N. Lee, Y.-J. Kim, and S. Kang (2004) Properties of self-assembled monolayer as an anti-adhesion layer on metallic nanomold. *The Nanotechnology Conference 2004* (Boston) 3, 350–353.

20. S. Y. Chou and P. R. Krauss (1997) Imprint lithography with sub-10 nm feature size and high throughput. *Microelectron. Engineering Journal* 35, 237–240.

21. S. Y. Chou, P. R. Krauss, and L. Kong (1996). Nanolithographically defined magnetic structures and quantum magnetic disk. *Journal of Applied Physics* 79, 6101–6106.

22. S. Lebib, Y. Chen, J. Bourneix, F. Carcenac, E. Cambril, L. Couraud, and H. Launois (1999) Nanoimprint lithography for a large area pattern replication. *Microelectronic Engineering Journal* 46, 319–322.

23. Y. Hirai, S. Harada, S. Isaka, M. Kobayshi, and Y. Tanaka (2002) Nanoimprint lithography using replicated mold by Ni electroplating. *Japanese Journal of Applied Physics* 41, 4186–4189.

24. N. F. Borrelli (1999) *Micro-Optics Technology*, Marcel Dekker, New York, 9–58.

25. S. Moon, N. Lee, and S. Kang (2003) Fabrication of a microlens array using microcompression molding with an electroformed mold insert. *Journal of Micromechanics and Microengineering* 13(1), 98–103.

26. Y. Kim, K. Sung, and S. Kang (2002) Effect of insulation layer on transcribability and birefringence distribution in optical disk substrate. *Optical Engineering* 49, 2276–2281.

27. C. Elsner, J. Dienelt, and D. Hirsch (2003) 3D-microstructure replication processes using UV-curable acrylates. *Microelectronic Engineering* 65, 163–170.

28. M. Grischke, A. Hieke, F. Morgenweck, and H. Dimigen (1998) Variation of the wettability of DLC-coatings by network modification using silicon and oxygen. *Diamond and Related Materials* 7, 454–458.

29. D.-Y. Wang, K.-W. Weng, C.-L. Chang, and W.-Y. Ho (1999) Synthesis of Cr_3C_2 coatings for tribological applications. *Surface and Coatings Technology* 120–121, 622–628.

30. D.-Y. Wang and M.-C. Chiu (2001) Characterization of Cr_2O_3/CrN duplex coatings for injection molding applications. *Surface and Coatings Technology* 137, 164–169.

31. K. Grundke, P. Uhlmann, T. Gietzelt, B. Redlich, and H-J. Jacobasch (1996) Studies on the wetting behaviour of polymer melts on solid surfaces using the Wilhelmy balance method. *Colloids and Surfaces A: Physicochemical and Engineering Aspects* 116, 93–104.

32. M. Wulf, S. Michel, K. Grundke, O. I. del Rio, D. Y. Kwok, and A. W. Neuman (1999) Simultaneous determination of surface tension and density of polymer melts using axisymmetric drop shape analysis. *Journal of Colloid and Interface Science* 210, 172–181.

33. S. Kang (2004) Replication technology for micro/nano-optical components. *Japanese Journal of Applied Physics* 43(8b), 5706–5716.

34. N. Lee, Y. Kim, and S. Kang (2004) Temperature dependence of anti-adhesion between a stamper with submicron patterns and the polymer in nanomoulding. *Journal of Physics D: Applied Physics* 37(12), 1624–1629.

35. K. Jung, H. Kim, N. Park, S. Kang, and Y. Park (2003) Design of the optical path for small form factor optical disk drive. *JSME-IIP/ASME-ISPS Joint Conference on Micromechatronics for Information and Precision Equipment, Japan*, 265–266.

36. N. Lee, Y. Kim, S. Kang, and J. Hong (2004) Fabrication of metallic nanostamper and replication of nanopatterned substrate for patterned media. *Nanotechnology* 15(8), 901–106.

CHAPTER 6

UV-Imprinting Process and Imprinted Micro/Nanostructures

6.1 INTRODUCTION

Polymer replication plays an important role in the fabrication of micro/nano-optical components at low cost and with high throughput. Two primary polymer replication methods have been used: the thermal method and the UV-curing method. In thermal imprinting, the resin is softened above the glass transition temperature and then shaped in a mold [1–5]. In UV imprinting, a transparent mold is applied to a liquid resin, which is then cured by exposure to UV light. Thermal imprinting requires a pressure of 10 MPa and significant throughput time, especially the time required for raising and reducing the temperature [6]. In contrast, UV imprinting does not require high pressures because the resin is basically a viscous liquid, soft enough to be easily deformed. However, as the resin is in liquid form and is polymerized during exposure, the quality of the imprinted optical components is sensitive to processing parameters such as curing temperature, curing time, intensity, and wavelength of the UV light, and compression pressure. Furthermore, problems caused by microair bubbles, shrinkage due to polymerization, and the nonuniformity of the residual layer are frequently encountered in UV-imprinted components.

This chapter covers designing and construction of a UV-imprinting system and process control. The photopolymerization process is described and implemented, and the degree of photopolymerization of the imprinted components is analyzed. The effects of the processing conditions on the elimination of microair bubbles, the replication quality, and the residual layer thickness are examined. Finally, some practical applications are shown.

Micro/Nano Replication: Processes and Applications, First Edition. Shinill Kang.
© 2012 John Wiley & Sons, Inc. Published 2012 by John Wiley & Sons, Inc.

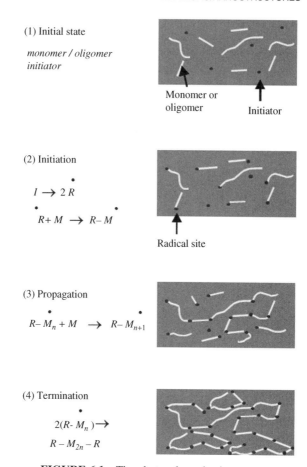

(1) Initial state

monomer / oligomer
initiator

Monomer or oligomer   Initiator

(2) Initiation

$$I \rightarrow 2\overset{\bullet}{R}$$

$$\overset{\bullet}{R} + M \rightarrow R{-}\overset{\bullet}{M}$$

Radical site

(3) Propagation

$$R{-}\overset{\bullet}{M}_n + M \rightarrow R{-}\overset{\bullet}{M}_{n+1}$$

(4) Termination

$$2(R{-}\overset{\bullet}{M}_n) \rightarrow$$
$$R - M_{2n} - R$$

FIGURE 6.1   The photopolymerization process.

6.2 PHOTOPOLYMERIZATION

UV photopolymerization is the solidification of a liquid photopolymer by exposure to UV light. The photopolymerization process can be divided into photoinitiation, propagation, and termination. Figure 6.1 illustrates the process [7,8]. A photopolymer consists of several elements: a photoinitiator, which absorbs photons to commence the reaction, a monomer/oligomer, and some additives for special purposes. When a photon is absorbed, the energy is transferred to the electronic structure of the photoinitiator. These results in one or more electronic transitions in the molecule. This excitation of the electrons, either directly (a type 1 photoinitiator) or by reaction with another molecule (a type 2 photoinitiator), ultimately leads to the formation of a free radical. Figure 6.2 shows the schemes for free radical formation from these photoinitiators [7].

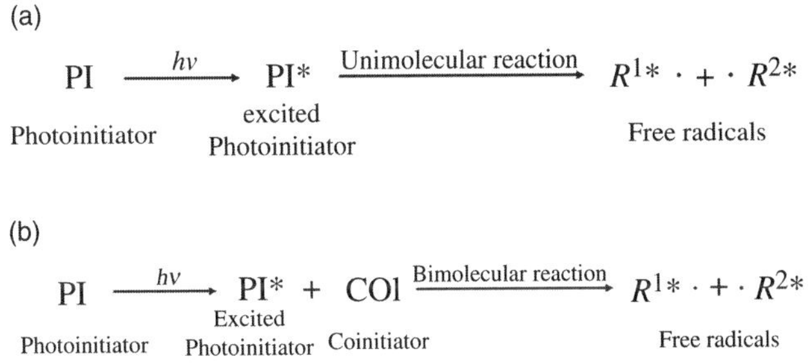

(a)

$$PI \xrightarrow{\ h\nu\ } PI* \xrightarrow{\text{Unimolecular reaction}} R^1* \cdot + \cdot R^2*$$

Photoinitiator   excited   Free radicals
Photoinitiator

(b)

$$PI \xrightarrow{\ h\nu\ } PI* + COI \xrightarrow{\text{Bimolecular reaction}} R^1* \cdot + \cdot R^2*$$

Photoinitiator   Excited   Free radicals
Photoinitiator Coinitiator

FIGURE 6.2   Formation of free radicals from photoinitiators: (a) type 1, which produces free radicals directly; (b) type 2, which reacts with another molecule, and ultimately results in the formation of a free radical.

The monomers and oligomers have double bonds that can easily be broken, so that the free radicals from the photoinitiator can react with them. Propagation is the repeated addition of monomeric units to the polymer backbone in a chain reaction. Chain transfer often occurs during the propagation period. Hydrogen abstraction by the radical on the growing polymer chain terminates the growing chain, with the concomitant production of a new radical. If the newly formed radical can start another chain reaction, then it is called a chain transfer process. The final stage of photopolymerization is termination. The chain reaction can be terminated by various processes, such as disproportion or a recombination reaction between two growing polymer chains. Termination can also occur upon recombination with other radicals, including the primary radicals produced by the photoreaction.

To obtain good mechanical and optical properties, as well as repeatability, the UV curing dose should be sufficient to produce a high degree of photopolymerization. In order to examine the effects of varying the exposure dose, the cured state of the photopolymer was analyzed with Fourier transform infrared (FTIR) spectroscopy. FTIR spectroscopy is an analytical technique that can be used to determine chemical structure by detecting the absorption or scattering of infrared light [6,9,10]. Most monomers for photopolymerization, including urethane acrylate, have carbon double bonds. During the polymerization of the photopolymer, the carbon double bond in the monomer state is converted into a carbon single bond, as depicted in Figure 6.3a [11]. In the monomer state, silicone urethane acrylate (which was used as a test material in this study) has absorption peaks in its FTIR spectrum at 3037 and 3100 cm^{-1} due to the carbon double bond, as shown in Figure 6.3b. However, these peaks disappear when the photopolymer is exposed to a UV dose exceeding 180 mJ/cm^2. An exposure dose of 180 mJ/cm^2 is sufficient for the complete polymerization of the photopolymer.

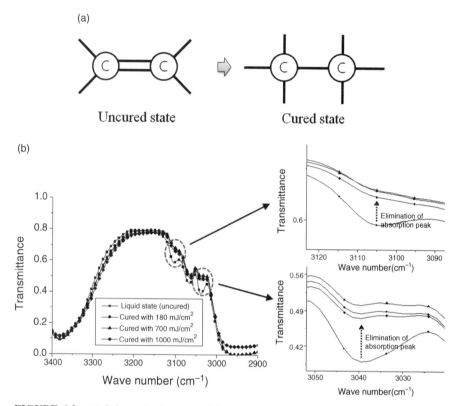

FIGURE 6.3 (a) Schematic diagram of the changes in carbon structure during the polymerization process; (b) The results of FTIR spectroscopy for the urethane acrylate resin. In the monomer state, silicone urethane acrylate has absorption peaks in its FTIR spectrum at 3037 and 3100 cm^{-1} due to its carbon double bonds. Reprinted with permission from Ref. [11]. Copyright 2006, American Institute of Physics.

6.3 DESIGN AND CONSTRUCTION OF UV-IMPRINTING SYSTEM

UV-imprinting system was designed and constructed as depicted in Figure 6.4 [12–14]. The UV-imprinting system consists of a UV exposure unit, a mold, and a substrate jig unit with precision mechanical stage, a microscope and monitoring unit, and a system for applying pressure from the bottom. The exposure unit is composed of a metal halide UV lamp with a wavelength range of 265–420 nm, an ellipsoidal reflector, a cold mirror, and a collimating lens. The imprinting jig unit is composed of a top glass, a vacuum chuck, a wedge compensation module, and a precision x, y, z, and θ stage. A transparent rubber layer with a thickness of 3 mm is attached to the top glass to ensure a uniform pressure distribution and to

(a)

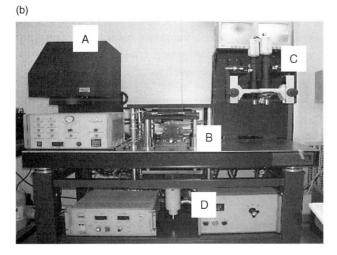

(b)

FIGURE 6.4   (a) Schematics and (b) picture of the UV-imprinting system that consist of [A] a UV-exposure unit, [B] a mold and substrate jig unit with a precision mechanical stage, [C] a microscope and monitoring unit, and [D] a system for applying pressure from the bottom.

hold a film-type mold. The exposure and microscope units are mounted on the dovetail rails to facilitate movement of the jig unit from the initial position. After the material coating and mold covering processes have been completed, the stacked substrate, material, and mold are placed on the vacuum chuck and brought into contact with the transparent rubber layer on the top glass by movement of the mechanical z stage. The top part of the stack (transparent mold or transparent substrate) is held together by the friction between it and the rubber layer, and thus, the mold and the substrate move independently during the alignment process. The microscope unit is moved on top of the jig unit, and the alignment process is then carried out. Following the alignment process, compression pressure is applied from the bottom of the wafer chuck to the top glass, and the microscope unit is returned to its initial position. The exposure unit is then moved to the top of the jig unit. Collimated UV light is irradiated on the replication material though the top glass, transparent rubber layer, and transparent mold or substrate. After the imprinting process is completed, the stacked substrate, cured photopolymer, and mold are removed from the imprinting system, and the releasing process is carried out. The UV-curing dose (intensity of UV light $\times$ curing time) is controlled with a mechanical shutter, and the compression pressure (with magnitude up to 200 kPa) is controlled with a hydraulic unit and a load cell.

6.4 UV-TRANSPARENT MOLD

In this chapter, to explain UV-transparent molds for UV-imprinting process, a specific example will be shown integrating a microlens array with precision alignment on an electronic device such as image sensor. Design, fabrication, and evaluation of UV-transparent mold will be presented in this chapter.

Integrating a microlens array with an electronic circuit by UV imprinting requires a mold that is transparent to UV light because the device side is opaque. In addition to being UV transparent, the mold must also have good releasing properties and high dimensional accuracy to obtain high replication quality and alignment accuracy. Fabricating a UV-transparent mold for a designed lens requires a master pattern having the same shape as the final imprinted lens. In this section, a microlens master was fabricated by photolithography and the reflow process. A base size of 4.65 by 4.65 μm and a height of 1.12 μm were obtained by spin coating the positive photoresist AZ1518 at 6000 rpm for 30 s, followed by a conventional photolithography process. The photoresist pedestal was reflowed at 160°C for 50 s on a hot plate. The sag height of the reflowed microlens was 1.426 μm, and the base size was about 4.63 by 4.63 μm. In this chapter, two types of UV-transparent mold, obtained by replication processes involving UV-curable silicone-urethane-acrylate-based photopolymer and PDMS, were fabricated from the same master pattern, and their properties were compared. PDMS is widely used as a mold material in soft molding because of its good UV transparency and releasing properties [15]. The silicone urethane acrylate UV-transparent mold was fabricated by replicating UV-curable silicone-urethane-acrylate-based photopolymer on an adhesion-layer-coated (polyethylene terephthalate (PET)) film. The UV radiation energy was 500 mJ/cm^2, and the compression pressure was 90 kPa. A PDMS mold was also fabricated by thermal curing at 40°C for 3 h. The characteristics of both UV-transparent molds, including the UV transmittance, releasing properties, and geometrical properties, were compared. The thickness of the fabricated silicone urethane acrylate UV-transparent mold sample was 270 μm, and that of the PDMS mold was 5 mm. The molds were exposed to a collimated UV light source with a wavelength of 365 nm, and the transmittance was measured using an intensity meter positioned under the molds. Both molds exhibited a transmittance of more than 90%, which is sufficient for UV imprinting. The releasing properties of both molds were evaluated by measuring the contact angles using the method described in Ref. [16]. A precise amount of photopolymer was dropped on the surface of each mold by a dispenser. After the photopolymer droplet was allowed to expand sufficiently, images were obtained with a CCD camera, and the contact angle was calculated. The contact angles were 62.5° for the silicone urethane acrylate mold and 68.1° for the PDMS mold, as depicted in Figure 6.5, which indicates that the two molds exhibited similar releasing properties.

In applications involving multilayer overlays, accurate alignment between substrate and mold is required. To accomplish this, in-plane shrinkage during the mold fabrication process should be eliminated. To analyze the in-plane shrinkage of UV-transparent molds, we fabricated a test mold using a microlens array master pattern with a designed array width of 2620.2 μm. Figure 6.6b shows the measured lateral length of the PDMS mold. The designed width of the microlens array was 2620.2 μm, but the width of the PDMS replicated mold was 2575 μm. This means that the lateral shrinkage of the PDMS mold was about 1.7%, while that of the silicone urethane acrylate mold was negligible, as shown in Figure 6.6a. The volumetric shrinkage of a photopolymer cannot be eliminated. In this case, however, the lateral shrinkage of the photopolymer was compensated by the vertical shrinkage of the residual layer because the backside film sustains the photopolymer. This implies the presence of stress, but no strain, on the silicone urethane acrylate.

In the UV-imprinting process, a relatively low pressure is usually applied to compensate for the shrinkage of the photopolymer. Therefore, a UV-transparent mold should have sufficient hardness to prevent the deformation

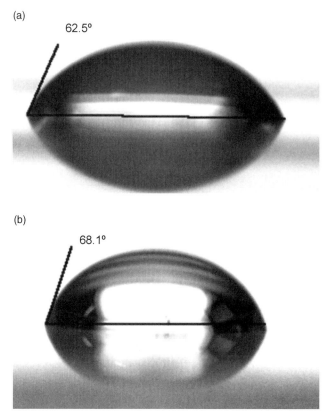

FIGURE 6.5 Images of the photopolymer droplets on (a) silicone urethane acrylate UV-transparent mold, (b) PDMS mold.

of the mold cavity. To evaluate the effect of applied pressure on the dimensional accuracy of a UV-transparent mold, UV imprinting was carried out under a range of applied pressures. Figure 6.7 shows the variation of the surface profiles of the mold cavity and the imprinted microlens with respect to the applied pressure using (a) a silicone urethane acrylate mold and (b) a PDMS mold. In the case of the silicone urethane acrylate mold, degradation of the replication quality due to material shrinkage can be compensated with applied pressure. In the case of the PDMS mold, however, the applied pressure cannot compensate for the shrinkage because the mold is deformed during the process. This means that silicone urethane acrylate molds are the preferred choice to compensate for photopolymer shrinkage.

The in-plane shrinkage and low hardness of the PDMS UV-transparent mold makes it unsuitable for the integration process, whereas the silicone acrylate UV-transparent mold has good UV transparency, releasing properties, and dimensional accuracy.

(a)

(b)

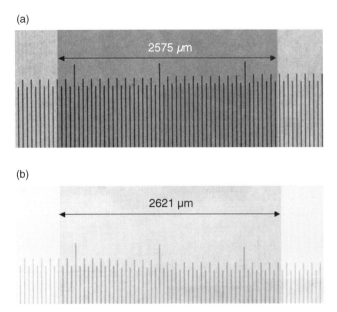

FIGURE 6.6   Microscope images of align marks during the alignment process between each mold and substrate; (a) Microscope image of microlens cavity array of PDMS mold. To evaluate lateral shrinkage of the PDMS mold, we replicated microlens array that has designed lateral length of 2620.2 μm, and the lateral length of the replicated PDMS mold was 2575 μm. It notes that the PDMS mold has about 1.7% lateral shrinkage, (b) Microscope image of microlens cavity array of flexible film mold. To evaluate lateral shrinkage of the flexible film mold, we replicated microlens array that has designed lateral length of 2620.2 μm, and the lateral length of the replicated flexible film mold was 2621 μm. It notes that the lateral shrinkage of flexible film mold is negligible.

6.5   EFFECTS OF PROCESSING CONDITIONS ON REPLICATION QUALITIES

Because of the high fluidity of the replication material, UV-imprinted optical components have high transcribability. However, the replication quality may be degraded by material shrinkage during the curing process [17,18]. It is important to maximize the replication quality of UV-imprinted parts to obtain micro/nano-optical components with desired properties. The degradation of replication quality in the UV-imprinting process can be prevented by controlling the material composition, mold shape, and processing conditions. Rudschuck et al. [19] fabricated a V-groove to examine the influence of various polymer formulations on replication quality. However, control of the material formulation can affect other optical and mechanical properties of imprinted parts, and the redesigning of mold cavities for micro/nano-optical

(a)

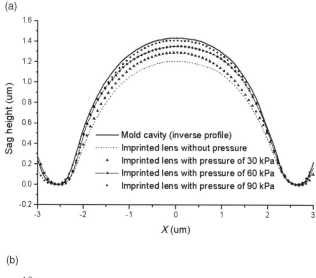

(b)

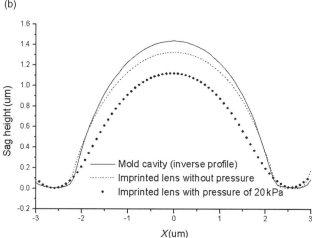

FIGURE 6.7 The variations of the surface profiles of the mold cavity and the UV- imprinted lens with applied pressure for (a) the silicone urethane acrylate UV-transparent mold and (b) the PDMS mold.

components is very difficult. In this section, the effect of processing conditions on the replication quality of an imprinted microlens array is examined.

To investigate the effect of processing conditions on replication quality, a microlens array was designed and fabricated. To fabricate a microlens array by UV imprinting, a mold containing lens-shaped microcavities is required. We fabricated a nickel mold insert for a microlens array with a lens diameter of 14 μm and a pitch of 15 μm via the electroforming of reflowed lenses. Figure 6.8 shows SEM images and three-dimensional surface profiles obtained from the AFM data of (a) the reflow master and (b) the

(a)

(b)

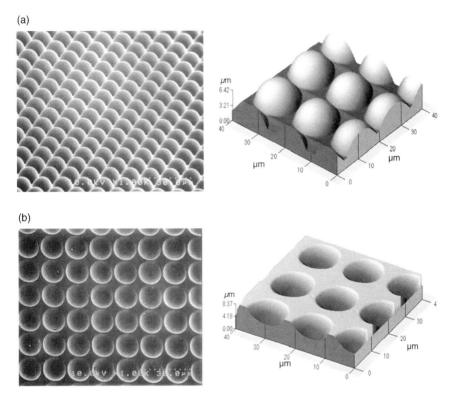

FIGURE 6.8   SEM images and three-dimensional-surface profiles obtained from AFM data of (a) the reflow master and (b) electroformed metal mold insert having microlens array with a lens diameter of 14 μm, a pitch of 15 μm. Reprinted with permission from Ref. [20]. Copyright 2003, Institute of Physics.

electroformed metal mold insert for the microlens array. The mean values of the sag height of the reflowed master and the depth of the mold cavity were 4.35 and 4.33 μm, respectively. The depth of the electroformed metal mold cavities deviated from the sag height of the reflow lenses by less than 0.04 μm in the center region. We believe this deviation was caused by AFM measurement errors on the various convex and concave curves.

We examined the effects of compression pressure and UV-curing dose on the replication quality of a microlens array with a lens diameter of 14 μm and a pitch of 15 μm. To quantify the replication quality, we measured the sag heights of the lenses using an AFM. Figure 6.9a shows the effect of compression pressure on the sag height of the imprinted microlens at a UV-curing dose of 400 mJ/cm^2. The sag height increased as the compression pressure increased, and its value approached that of the reflow lens master (4.358 μm) at a compression pressure of 90 kPa or higher because the shrinkage of the UV-curable photopolymer was compensated by the applied

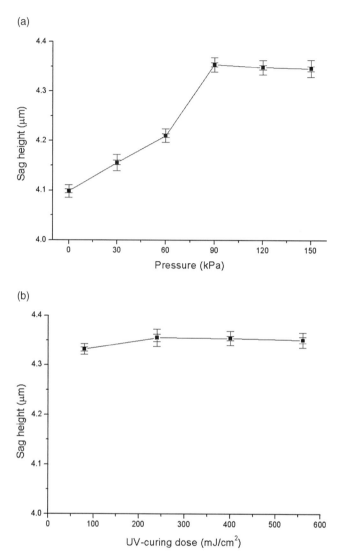

FIGURE 6.9 Sag heights of imprinted microlens for increasing values of (a) compression pressure with UV-curing dose fixed at 400 mJ/cm^2 and (b) UV-curing dose with compression pressure fixed at 90 kPa. Reprinted with permission from Ref. [20]. Copyright 2003, Institute of Physics.

pressure. However, excessive compression pressure can cause internal defects such as microcracks, depending on the curing level. Figure 6.9b shows the effect of UV-curing dose on the sag height of the imprinted microlens at a compression pressure of 90 kPa. The UV-curing dose appears to have no significant effect on the sag height. Figure 6.10a compares the surface profile of the reflow lens array with those of corresponding imprinted

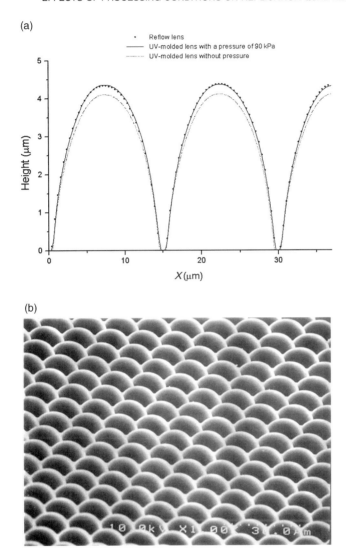

FIGURE 6.10   (a) Comparison between surface profiles of mother lens (reflow lens) and those of imprinted lens arrays with pressure of 90 kPa and without pressure and (b) SEM image of UV-imprinted microlens array with a diameter of 14 μm and a pitch of 15 μm, which was fabricated in optimal process conditions, pressure of 90 kPa and UV-curing dose of 400 mJ/cm^2. Reprinted with permission from Ref. [20]. Copyright 2003, Institute of Physics.

lens arrays at pressures of 90 and 0 kPa. The deviations between the sag heights of the reflow lens and the UV-imprinted lens, with and without an applied pressure of 90 kPa, were 0.01 and 0.26 μm, respectively. The degradation of replication quality by shrinkage was 5.9% in the microlens with a diameter of 14 μm, but was improved to less than 0.3% by applying a pressure

(a)

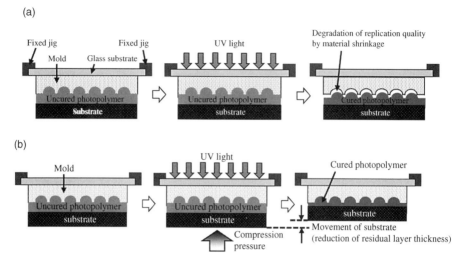

(b)

FIGURE 6.11   Comparison of schematics of UV-imprinting process between without and with compression pressure: (a) degradation of shape accuracy in UV imprinting without compression pressure; (b) compensation of material shrinkage in the lens cavity by compression pressure.

of 90 kPa. The repeatability errors in these experiments were less than 0.02 μm. (In other words, these results had repeatability.) Figure 6.10b shows an SEM image of a UV-imprinted microlens array with a lens diameter of 14 μm and a pitch of 15 μm, fabricated under optimal process conditions (pressure of 90 kPa and UV-curing dose of 400 mJ/cm^2).

These results show that the degradation of shape accuracy can be compensated by compression pressure during the curing process. The mechanism of compensation is explained schematically in Figure 6.11. Figure 6.11a shows the deterioration of shape accuracy caused by material shrinkage in the UV-imprinting process without compression pressure. Figure 6.11b shows the compensation mechanism of material shrinkage in the UV-imprinting process with applied pressure [20,21]. It should be noted that in the UV-imprinting process with compression pressure, material shrinkage during the curing process occurred primarily in the residual layer, rather than in the lens cavity.

6.6   CONTROLLING OF RESIDUAL LAYER THICKNESS USING DROP AND PRESSING METHOD

Another important issue in the UV-imprinting process is control of the distance between the microlens array and the device (residual layer thickness). A spin coating process can be used to control this thickness and can provide a thin residual layer with high uniformity [22]. However, it is a difficult and time-consuming procedure to remove the microair bubbles that are often generated on uniformly coated substrates during the mold-covering process. To prevent the formation of microair bubbles, the material is dropped onto the middle of the substrate and spread out by compression. In this technique, the thickness of the residual layer (i.e., the remaining layer under the lens base) determines the distance between the device and the microlens, and this distance should be controlled to obtain the desired optical properties in the integrated optics. The thickness of the residual layer can be affected by the viscosity, the applied pressure, the amount of material in the initial drop, and the UV-curing dose. In this chapter, to determine the optimum processing conditions that promote the formation of a residual layer with uniform thickness on a 4 inch wafer, we experimentally investigated the effects of varying the applied pressure, the amount of material in the initial drop, and the UV-curing dose on residual layer thickness. A UV-curable urethane-acrylate-based photopolymer with a viscosity of 300 cps was used in this research. Figure 6.12a shows the effect of applied pressure on the thickness of the residual layer for a fixed UV-curing dose of 500 mJ/cm^2 and an initial drop of mass 0.62 g. The mean value of the residual layer thickness decreased as the applied pressure was increased. However, the thickness variation was minimized at an applied pressure of 90 kPa. Figure 6.12b shows the residual layer thickness distributions on the 4 inch wafer for various applied pressures. When compression pressure is applied, the initial drop expands from the center to the outside of the wafer. At low applied pressures, the drop did not expand sufficiently. At high applied pressures, however, the material in the outer regions expanded beyond the substrate. Figure 6.13 shows (a) the effect of the initial drop amount on the residual layer thickness for a fixed UV-curing dose of 500 mJ/cm^2 and a pressure of 90 kPa and (b) the effect of UV-curing dose for a fixed initial drop mass of 0.62 g and a pressure of 90 kPa. These results show that the amount of material in the initial drop and the UV-curing dose do not affect the residual layer thickness. We conclude that a compression pressure of 90 kPa should be applied to obtain uniform residual layer thickness. The optimum residual layer thickness for the present UV-curable photopolymer was found to be 20 μm. Even though these results are specific to this particular type of UV-curable photopolymer, the same experimental approach can be applied to other UV-curable materials. This material-coating technique was utilized to replicate the UV-transparent mold from the reflowed master in this study.

6.7 ELIMINATION OF MICROAIR BUBBLES

Microair bubbles or air inclusion inside the photopolymer or at the interface between the photopolymer and the mold are frequently observed

(a)

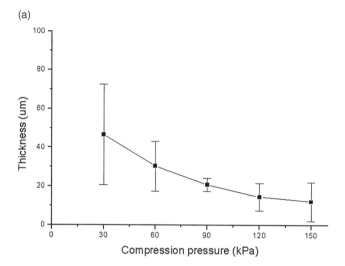

(b)

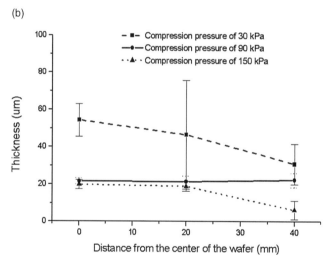

FIGURE 6.12 Effects of applying compression pressure to the urethane acrylate UV-curable photopolymer on the residual layer thickness: (a) effects of varying the thickness distribution for various applied pressures for a fixed UV-curing dose of 500 mJ/cm^2 and an initial drop of 0.62 g.

in UV-imprinted parts, and they can cause various defects [15,21]. One can use evacuation to avoid the formation of such bubbles, but this entails a high initial cost and a long cycle time. The microair bubble problem is caused by air bubble inclusion in the initial liquid photopolymer and the generation of air bubbles during the material-coating and mold-covering processes. Air bubble inclusion in the initial replication material can easily be eliminated by applying a degassing procedure to the initial photopolymer in

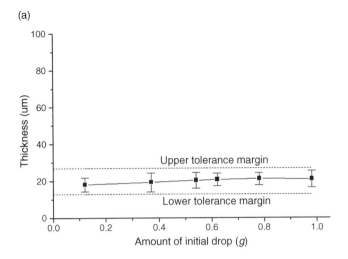

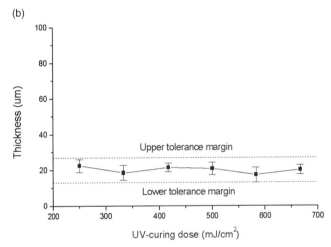

FIGURE 6.13   Effects of varying (a) the amount of material in the initial drop (for a fixed UV-curing dose of 500 mJ/cm^2) and (b) the UV-curing dose on the residual layer thickness (for a fixed compression pressure of 90 kPa and a fixed initial drop of 0.62 g).

a vacuum chamber. Air bubbles at the photopolymer–mold interface that develop during the material-coating or mold-covering processes can be eliminated by preheating. We heated the photopolymer to a temperature of around 40°C, at which point the viscosity of the photopolymer decreases. After the removal of air from the photopolymer coated on the substrate, the mold was carefully covered with the material to prevent the formation of air bubbles. Figures 6.14a and b show images of pyramidal pits and microlenses with air bubbles. Figures 6.14c and d show images of pyramidal pits and microlenses after the removal of microair bubbles. The mold insert for the pyramidal pits

(a)

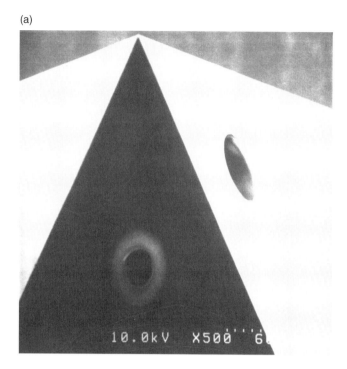

(b)

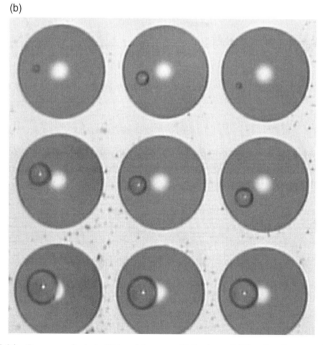

FIGURE 6.14 Images of air bubbles (a) pyramidal pit and (b) microlenses with microair bubbles, and the images of (c) pyramidal pits and (d) microlenses after removal of microair bubbles. Reprinted with permission from Ref. [21]. Copyright 2003. SPIE.

(c)

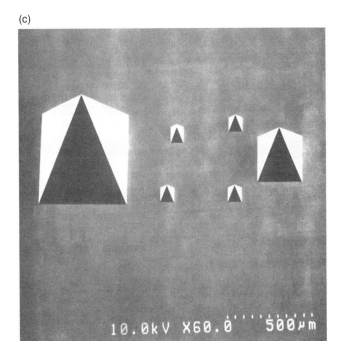

(d)

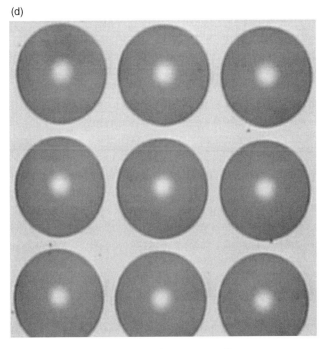

FIGURE 6.14   *(Continued)*

was fabricated by conventional anisotropic wet etching of a (100) silicon wafer, and the mold insert for the microlens array was fabricated via the reflow method and microelectroforming.

6.8 APPLICATIONS

6.8.1 Wafer Scale UV Imprinting

UV imprinting is suitable for integrating micro/nano-optics with electronic circuits, as its processing conditions are compatible with electronic circuit fabrication processes, and it produces micro/nano-optical components with low thermal expansion, enhanced stability, and low birefringence. The integration process should be carried out on the wafer scale because electronic devices are fabricated on the wafer scale, and wafer-scale integration processing enables the mass production of micro/nano-optics with high alignment accuracy and uniformity. The main issues in wafer-scale UV imprinting are the shape accuracy and uniformity of the integrated optics, the alignment accuracy, and the uniformity of the residual layer, as noted in the previous chapter.

Integrating micro/nano-optics with electronic circuits via UV imprinting requires the mold to be transparent to UV light, as the device side is opaque. Good releasing properties and high dimensional accuracy are also required in the UV-transparent mold to obtain high replication quality and alignment accuracy in the wafer-scale integration process. In this study, two types of UV-transparent mold were fabricated, and their properties were compared. Wafer-scale UV imprinting of a microlens array was performed on an even surface, and the shape accuracy and uniformity, alignment accuracy, and uniformity of the residual layer thickness were analyzed [22].

Wafer-scale UV imprinting was carried out on a silicon substrate with an even surface, using a silicone urethane acrylate UV-transparent mold replicated from a reflowed lens master with a lens base size of 4.63 by 4.63 μm and a sag height of 1.426 μm. A liquid urethane-acrylate-based UV-curable photopolymer was used as the lens material. The photopolymer was spin coated on the substrate at 6000 rpm for 30 s to reduce the thickness of the residual layer and increase the thickness uniformity. After covering the mold to prevent the formation of microscopic air bubbles and completing the optical alignment process, UV imprinting was carried out at an irradiation energy of 2000 mJ/cm^2. Material in-plane shrinkage during the curing part of the process was compensated by the adhesion force between the photopolymer and substrate, as in the silicon acrylate UV-transparent mold fabrication process described in section 6.4. Out-of-plane shrinkage can be compensated by applying a compression force to the stack consisting of the substrate, the photopolymer, and the mold. Figure 6.15

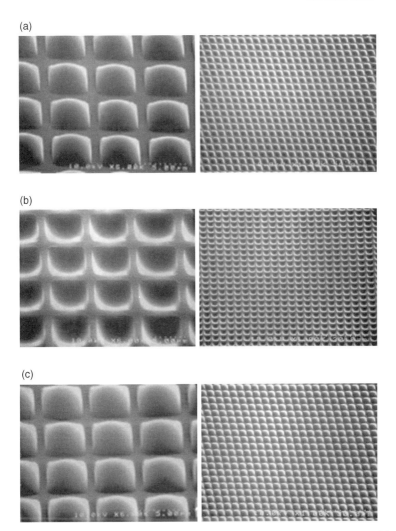

FIGURE 6.15 SEM images of (a) reflowed master, (b) UV-transparent mold, and (c) UV imprinted microlens array with a base size of 4.63 by 4.63 μm. Reprinted with permission from Ref. [22]. Copyright 2006. Optical Society of America.

shows SEM images of (a) the reflowed master and (b) the UV-transparent mold and UV-imprinted microlens array with base size of 4.63 by 4.63 μm. A microlens array with a side length of 4.63 μm, a sag height of 1.416 μm, a radius of curvature of 2.544 μm, and a surface roughness ($Rrms$) of 1.07 nm was fabricated from UV-curable urethane acrylate material. Deviations between master lens and integrated lens are shown in Figure 6.16.

Table 6.1 summarizes the geometrical properties of the designed lens, mold cavity, and imprinted microlens array. A microscope was used to measure the lens side length, and an atomic force microscope was used to

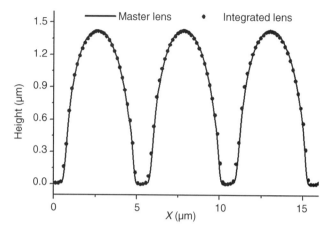

FIGURE 6.16 Comparison of surface profiles of the master and integrated microlens arrays with a side length of 4.63 µm and a pitch of 5.2 µm. The sag heights of the master and integrated lenses were 1.431 and 1.418 µm, respectively.

TABLE 6.1 Summary of the Geometrical Properties of the Designed and Integrated Microlens Arrays

Parameter		Value					
		Designed Lens		Mold Cavity		Imprinted Lens	
Surface Roughness (nm)	R_{p-v}	< 25.0		10.21		11.1	
	R_{rms}	< 2.5		1.05		1.07	
	R_a	< 1.8		0.79		0.81	
Side length (µm)	L_d	4.65	L_m	4.62	L_m	4.634	
Sag height (µm)	H_d	1.43	H_m	1.38	H_m	1.416	
Radius of curvature (µm)	R_d	2.51	R_m	2.22	R_m	2.544	
$|(L_d-L_m)/L_d| \times 100$		–		0.64%		0.34%	
$|(H_d-H_m)/H_d| \times 100$		–		3.49%		0.98%	
$|(R_d-R_m)/R_d| \times 100$		–		11.5%		1.35%	

measure the sag height. Because of the difference in AFM measuring conditions for convex and concave surfaces, the mold cavity deviates to some extent from the designed lens. The differences in the side lengths, sag heights, and radii of curvature of the designed and imprinted lenses were 0.34, 0.98, and 1.35%, respectively. Figure 4.5 compares the surface profiles of the reflowed master and the imprinted microlens array. The maximum deviation between designed and measured values was less than 0.015 μm.

To increase the alignment accuracy in the UV-imprinting process, alignment was carried out with a circular grating that produces a moiré effect, in addition to the conventional cross-patterns. The accuracy of the conventional alignment process is limited by the resolution of the microscope. Use of a circular grating increases the accuracy, as a small amount of misalignment results in a large-pitch moiré fringe pattern that can easily be detected, even by a low-resolution imaging system [23,24].

In the wafer-scale integration of a microlens array, uniformity and repeatability are very important. Not only the shape of the array, but also the thickness of the residual layer is critical because in the wafer-scale integration of a microlens array on an image sensor, the residual layer replaces the planarization layer, and its thickness affects the optical path. To test the uniformity and repeatability of the imprinted microlens array in wafer scale, the sag height and thickness of the residual layer were measured using an atomic force microscope and a prism coupler thickness-measuring system. We measured the sag heights of 12 sample lenses in each array, four arrays on each wafer, and three wafers processed under the same processing conditions. Table 6.2 lists the results. The standard deviation of the sag heights was 0.024 μm for the same chip, 0.038 μm for the same wafer, and 0.037 μm for the same chip position on different wafers. We also measured the thickness of the residual layer at 21 sampling points on three wafers. Table 6.3 lists the results. The standard deviation of the thickness of the residual layer was 0.164 μm. From the statistical analysis of the results, we can infer that the process had uniformity and repeatability on a wafer scale.

6.8.2   Diffractive Optical Element

The demand for small, high-capacity optical data storage devices has increased rapidly. The areal density of an optical disk can be increased by using an objective lens with a higher numerical aperture and a light source with a shorter wavelength. Several types of objective lenses have been proposed for small-form-factor optical drives [25]. Because the objective lenses for such drives must be held to tight assembly tolerances, wafer stacking offers a suitable fabrication process. We have designed and fabricated a wafer-scale stacked micro-objective lens with a numerical aperture of

TABLE 6.2 Mean Sag Heights and Deviations of Imprinted Microlens Arrays

		Measured Array				A, B, C and D
		A	B	C	D	
Wafer 1	$H_m(\mu m)$	1.460	1.438	1.359	1.400	1.414
	$\sigma(\mu m)$	0.018	0.015	0.015	0.022	0.038
Wafer 2	$H_m(\mu m)$	1.468	1.455	1.381	1.432	1.434
	$\sigma(\mu m)$	0.019	0.024	0.017	0.019	0.033
Wafer 3	$H_m(\mu m)$	1.429	1.421	1.362	1.391	1.401
	$\sigma(\mu m)$	0.016	0.017	0.020	0.016	0.028
Wafer 1, 2, and 3	$H_m(\mu m)$	1.452	1.438	1.367	1.408	1.416
	$\sigma(\mu m)$	0.030	0.037	0.022	0.018	0.032

0.85 and a focal length of 0.467 mm for a 405 nm blue–violet laser. Its specifications are given in Table 6.4. A diffractive optical element (DOE) was used to compensate for the spherical aberration of the objective lens.

In this chapter, an eight-stepped DOE pattern master was fabricated by photolithography and reactive ion etching of a silicon substrate. A UV-transparent mold with good releasing properties and no lateral shrinkage was replicated using silicon acrylate UV photopolymer on a polyethylene tere-phthalate (PET) film. To find the optimum process conditions for UV imprinting, the degree of polymerization of the photopolymer and the effect of applied pressure were evaluated. Finally, the surface profile and diffraction efficiency of the imprinted DOE were analyzed.

To replicate a DOE for a wafer-scale stacked micro-objective lens, a master pattern with the same shape as the final DOE pattern was first required. The master pattern for the DOE was fabricated via photolithography and reactive ion etching on a silicon wafer. To generate an eight-stepped DOE, a sequence of photolithography and dry-etching processes was repeated three times using three

TABLE 6.3 Summary of Measured Thickness of Residual Layer

Measured Position	Thickness (μm)		
	Wafer 1	Wafer 2	Wafer 3
A	0.96	1.19	1.00
B	1.40	1.17	1.28
C	1.08	0.96	1.00
D	0.89	1.18	1.34
E	1.31	1.41	1.20
F	0.90	1.16	1.23
G	1.09	1.06	1.38
t_a	1.09	1.16	1.20
σ	0.184	0.128	0.141
$t_{p\text{-}v}$	0.51	0.45	0.34
t_a	1.152		
σ	0.164		
$t_{p\text{-}v}$	0.52		

TABLE 6.4 Design Specification of Assembled Wafer-Scale Stacked mIcro-Objective Lens

Wavelength (λ)	408($-5/+10$) nm
Focal length (f)	0.467 nm
Numerical aperture (NA)	0.85
Entrance pupil diameter (P_E)	φ0.8 mm
Working distance (d)	40 um
Cover layer	0.1 mm ($n_d = 1.58$)
Overall length (L)	1.36 mm
Wavefront error (λ_{rms})	0.0078 λ_{rms}

different masks. To produce the minimum critical dimension of 1.22 µm and a variety of DOE patterns on the wafer, we used a stepper with a projection ratio of 5:1. Each of the processes used to create the staircase pattern on the substrate included resist coating, mask alignment, mask pattern exposure, developing, and dry etching [26,27]. The total etching depth of the DOE was 605 nm. Twenty-one DOE patterns were generated on a silicon substrate. Figure 6.17 shows SEM

(a)

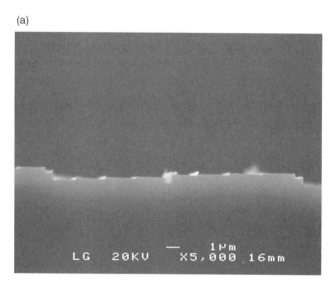

(b)

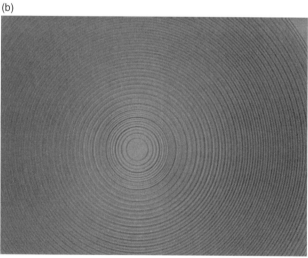

FIGURE 6.17 SEM images of master pattern of DOE. (There are some defects in the master pattern caused by misalignment of stepping process. We used a stepper with 0.2 µm alignment error and projection ratio of 5:1, so the defect of master pattern was created by this error.) Reprinted with permission from Ref. [12]. Copyright 2006. Institute of physics Science.

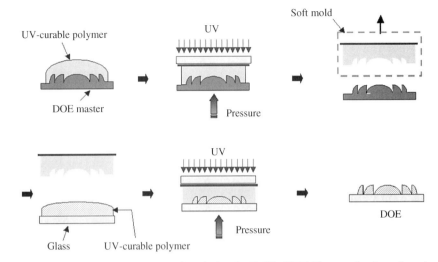

FIGURE 6.18 Process flow of UV imprinting for DOE. (PET film was laminated on the DOE master pattern that had been coated with UV-curable photopolymer. Mold was released after UV light was projected through the PET film for photopolymerization of mold material with applying pressure. Using the fabricated mold, DOE replica was fabricated on glass substrate.)

images of the master DOE pattern. Some defects in the master pattern occurred, which was caused by misalignment of the stepping process. We used a stepper with an alignment error of 0.2 μm, and the defects in the master pattern were created by this error.

UV imprinting was used to replicate the DOE pattern. Figure 6.18 illustrates the process flow [28,29]. A UV-imprinting system was constructed to carry out the process. The system consists of a UV-exposure system, an ultraprecision stage, a microscope and monitoring system, and a system for applying pressure from the bottom.

A silicon-urethane-acrylate-based photopolymer was selected as an imprinting material because it has good optical and mechanical properties. Fourier transform infrared spectroscopy (FTIR) was used to determine the proper exposure time.

During the UV-imprinting process, surface wrinkling occurred frequently, resulting in reduced diffraction efficiency of the DOE. This wrinkling was caused by the fact that the surface in contact with the UV-transparent mold solidifies more rapidly than the opposite side. The effect of applied pressure on the surface roughness was examined. It is clear from Figure 6.19 that the surface-wrinkling problem can be eliminated by applying sufficient pressure. However, excessive applied pressure can cause degradation of the surface quality. Excessive pressure may cause the photopolymer to stick to the surface. Figure 6.20 shows an SEM image of a DOE replicated at an applied pressure of 50 kPa.

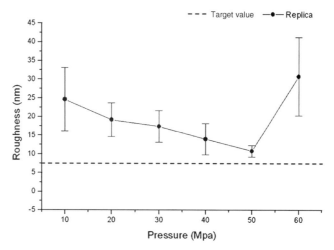

FIGURE 6.19 Surface roughness of master, mold, and replica for increasing values of applied pressure. (The surface-wrinkling problem can be eliminated by sufficient applied pressure, but excessive applied pressure can cause degradation of surface quality. Excessive pressure may result in stiction of photopolymer at the surface.) Reprinted with permission from Ref. [12]. Copyright 2006. Institute of physics Science.

FIGURE 6.20   SEM image of replicated DOE. (The replicated DOE that was fabricated with applied pressure of 50 kPa.) Reprinted with permission from Ref. [12]. Copyright 2006. Institute of physics Science.

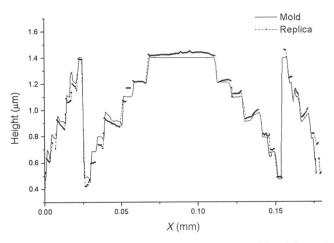

FIGURE 6.21 Comparison of surface profiles between the mold and the replicated DOE. (The deviation between the mold and the imprinted DOE was less than 0.1 μm.) Reprinted with permission from Ref. [12]. Copyright 2006. Institute of physics Science.

The geometrical properties and diffraction efficiency of the imprinted DOE were measured to evaluate the replication process. Figure 6.21 compares the surface profiles of the designed and replicated DOEs. The deviation between the mold and the imprinted DOE was less than 0.1 μm.

Diffraction efficiency is defined as the ratio of the intensity of the incident light to the intensity of the first-order diffracted light, expressed by the following:

$$\eta_{efficiency} = \frac{I_{DOE}}{I_{reference}} \times 100(\%) \tag{1}$$

The diffraction efficiency of the replica was measured by a diffraction-efficiency measuring system consisting of a blue–violet laser diode with a wavelength of 405 nm, a sample holder, an aperture, and an optical power meter. The intensity of the reference light was measured by detecting the intensity of light passing through a dummy substrate of exactly the same material and thickness as the DOE. Figure 6.22a shows a schematic of the measurement system for the intensity of the reference light. To measure the intensity of the first-order diffraction, we substituted the DOE for the dummy substrate. A 500 μm aperture was mounted at the first focusing distance to filter the first-order diffracted beam. Figure 6.22b shows a schematic of the measurement system for the intensity of the first-order diffracted light. Table 6.5 lists the measurements and the calculated efficiencies. The total average diffraction efficiency of the first-order diffraction was 84.5%.

(a)

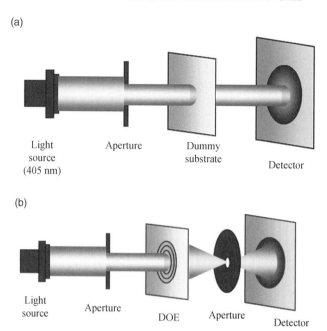

(b)

FIGURE 6.22 (a) Measurement system for intensity of reference light, (b) measurement system for intensity of first-order diffraction light. (The intensity of reference light was measured by the following process. Detecting the intensity of light through the dummy substrate that is of exactly the same material and thickness as the DOE. To measure the intensity of the first-order diffraction, we substituted DOE for dummy substrate. The 500 μm aperture was mounted at the first focusing distance for filtering first-order diffracted beam.) Reprinted with permission from Ref. [12]. Copyright 2006. Institute of physics Science.

6.8.3 Roll-to-Roll Imprint Lithography Process

Flexible electronic circuits are now a key component in the fields of flexible solar cells, flexible displays, wearable computers, radio frequency identification, portable information devices, and so on. Since the use of flexible electronic circuits has increased rapidly, the establishment of a low-cost method for patterning conductive patterns and tracks has become a priority. In conventional flexible electronics fabrication processes, the conductive tracks are formed by lift-off or etching the deposited metallic layer using an etch mask formed by photolithography or ink printing [30–33]. As discussed above, conventional track patterning processes are very complex and expensive when fabricating large areas of electronic circuits. Conventional imprint lithography is an emerging technology that enables low-cost and high-throughput nanofabrication. Unfortunately, this process is discontinuous and it is difficult to process large-scale substrates. However, the UV-roll imprint lithography process allows fabrication to be carried out continuously, with

TABLE 6.5   The Efficiency of the Replicated DOE

Sample	$I_{reference}$ (μW)	I_{DOE} (μW)	Efficiency (%)
Sample 1	2.67	2.23	83.52
		2.24	83.90
		2.32	86.89
		Average	84.77
Sample 2	2.63	2.22	84.41
		2.23	84.79
		2.25	85.55
		Average	84.92
Sample 3	2.71	2.31	85.24
		2.31	85.24
		2.29	84.50
		Average	84.99
Sample 4	2.71	2.35	83.93
		2.34	83.57
		2.30	82.14
		Average	83.21

high-precision micronano patterns on large area substrates [11,34]. Since it is a continuous process it has the advantages of cost effectiveness and high-throughput processing. Recently, Guo et al. and Chou et al. developed the roll-to-roll imprint lithography process and showed the possibility of mass production with micro- to nanoprecision patterning [35,36]. In our previous study [11], we showed the feasibility of constructing various polymer nanostructures on large area flexible substrates using a UV-roll imprint lithography process. In this study, a complete lithography technique is demonstrated to obtain a conductive indium tin oxide (ITO) pattern on the flexible substrate. We developed a lithography method that used roll-imprinted patterns as an etch mask for the lithography process. Here, a photopolymer pattern was imprinted on an ITO-coated polymer film using a continuous UV-roll imprinting process and the substrate was cut before subsequent processing. The residual layer was removed by a reactive ion etching (RIE) process, and ITO conductive patterns and tracks were formed by a wet-etching process. As practical examples, various conductive patterns and tracks with line widths of 2–20 μm were designed and fabricated. The present method was verified by analyzing the geometrical and electrical properties of the fabricated conductive patterns and tracks.

Among the various candidates for roll stamps for a UV-roll imprinting process, we propose metallic roll stamps in this study because they offer

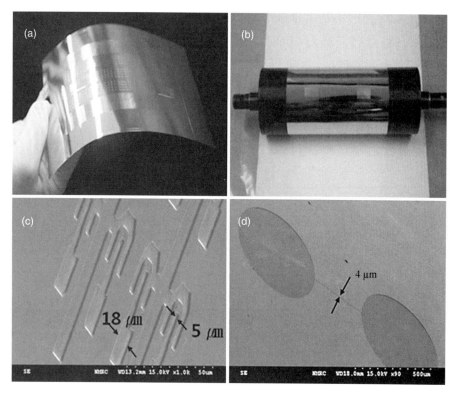

FIGURE 6.23 (a) Replicated nickel stamp prepared by electroforming (nominal thickness 100 μm); (b) metallic roll stamp (roll diameter 50 mm, length 210 mm); (c) SEM image of LCD patterns in the roll stamp; (d) SEM image of track with contact pads in the roll stamp (line length 300 μm, line width 4 μm, pad diameter 500 μm). Reprinted with permission from Ref. [38]. Copyright 2009, Institute of Physics.

durability for possible mass production, dimensional accuracy, and pattern fidelity [37]. To fabricate a roll stamp that contains various patterns and tracks with line widths of 2–20 μm, a thin nickel stamp, as shown in Figure 6.23a, was first replicated by an electroforming process from master patterns that were fabricated by photolithography.

After the electroforming process, a stamp surface roughness Rms of 50 nm was achieved by a back-polishing process. To increase the releasing properties of the roll stamp, anti-adhesion coating of the nickel stamp was required. In order to apply a self-assembled monolayer (SAM) of dichlorodimethylsilane (DDMS) as an anti-adhesion layer on the nickel stamp by immersion, a interfacial layer of SiO_2 with 10 nm thickness was prepared by plasma-enhanced chemical vapor deposition on the nickel stamp, which was active to the head group of DDMS. The contact angle of the nickel stamp increased from 60–114° after SAM treatment with DDMS. The electroformed thin metal

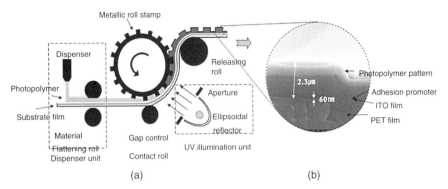

FIGURE 6.24 (a) Schematic diagram of roll-imprint lithography system. (b) Cross-sectional SEM image of replicated patterns on an ITO-coated PET substrate. The replicated pattern structure consists of the PET film, the ITO layer, the adhesion promoter layer, and the photopolymer pattern. Reprinted with permission from Ref. [38]. Copyright 2009, Institute of Physics.

stamp with a thickness of 100 μm was cut to the right size and then wrapped around the aluminum roll base using engineering glue. The discontinuity in the joint was used as a cutting line of the substrate. Figure 6.23b, shows the metallic roll stamp with a diameter of 50 mm and a length of 210 mm, and Figures 6.23c and d show SEM images of the liquid crystal display (LCD) patterns and tracks in the metallic roll stamp.

To fabricate high-quality large-scale conductive patterns and tracks, a continuous UV-roll imprint lithography system was used as explained in Ref. [11]. Figure 6.24a shows a schematic drawing of the UV-roll imprinting system. The UV-roll imprinting system consists of a dispensing unit, an imprinting unit, and a separating unit. First, at the dispensing unit, photopolymer was coated onto the ITO film by using the dispenser and the coated photopolymer was uniformly distributed by the flattening rolls. Then, at the imprinting unit, the contact roll for gap control applied a compressive force onto the substrate and the roll stamp. UV light was exposed for polymerization. Finally, the releasing roll unit separated replica and roll stamp at the separating unit.

For the UV roll imprint lithography process, an ITO layer of 60 nm thickness was first coated on the PET substrate with a thickness of 125 μm by low temperature sputtering and then an adhesion promoter was coated on the ITO layer. As an etch mask for lithography, acrylate-based photopolymer patterns were replicated on the ITO layer by UV-roll imprinting and the substrate was cut at the joint line of the roll stamp. Figure 6.24b shows the cross-sectional SEM image of replicated patterns on the ITO-coated PET substrate. It should be noted that preheating at 60–70°C was applied to the photopolymer to reduce the viscosity of the resin and to eliminate air bubbles within the resin. A reduced

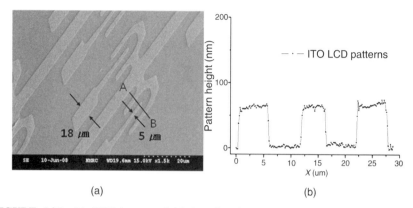

(a)   (b)

FIGURE 6.25 (a) SEM images of fabricated ITO conductive LCD patterns, (b) cross-sectional profile from A to B shown in Figure 6.25a measured by AFM. Reprinted with permission from Ref. [38]. Copyright 2009, Institute of Physics.

viscosity of the resin allowed a thinner residual layer. Normal thickness of the residual layer after the roll-imprinting process was 1 μm. In order to remove the residual layer, an RIE process was carried out. After residual layer removal the exposed ITO layer was etched by wet etching using the polymer patterns as etch mask. LCE12K was mixed with DI water with a volumetric ratio of 50:50, which was used as ITO etchant [39]. Finally, 5% of NaOH aqueous solution was used to remove the polymer etch mask [40].

Figure 6.25a shows an SEM image of fabricated ITO conductive LCD patterns. Figure 6.25b shows the cross-sectional profile from A to B shown in Figure 6.25a measured by atomic force microscopy (AFM). To examine the electrical property, a conductive track with contact pads was designed and

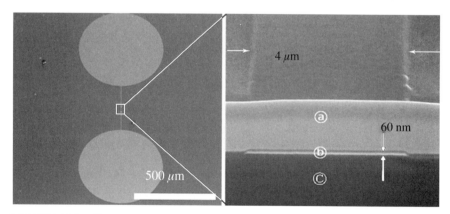

FIGURE 6.26 SEM image of conductive track with contact pads for the case of a line length of 300 μm and line width of 4 μm, cross-sectional SEM image of the conductive track of 4 μm width after FIB milling: (a) Pt layer to protect ITO layer for FIB milling, (b) ITO layer of conductive patterns with a thickness of 60 nm, (c) PET film region. Reprinted with permission from Ref. [38]. Copyright 2009, Institute of Physics.

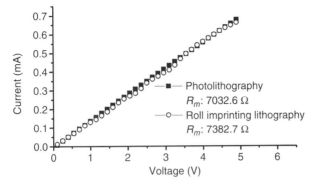

FIGURE 6.27 Comparison of electric resistance between photolithography processed ITO conductive tracks and those prepared by the present method. Reprinted with permission from Ref. [38]. Copyright 2009, Institute of Physics.

fabricated by the present roll-imprint lithography process. Figure 6.26a shows an SEM image of the conductive track with contact pads for the case of a line length of 300 µm and line width of 4 µm. The ITO layer structure was analyzed using cross-sectional SEM images as shown in Figure 6.26b, where focused ion beam (FIB) milling was used to prepare the cross-section of the ITO layer. Layer ⓑ shows the patterned ITO layer with a thickness of 60 nm, layer ⓐ is Pt to protect the ITO patterns during the FIB milling, and layer ⓒ is the PET substrate. The electrical resistance of the ITO track fabricated by roll-imprint lithography was measured and compared with that prepared by photolithography. Figure 6.27 shows the measured current (*mA*) as a function of voltage (*V*) of the ITO track fabricated by photolithography and the present method. The average electrical resistance of the ITO track for the case of a line length of 300 µm and a line width of 4 µm was 7032.6 Ω for photolithography and 7382.7 Ω for the present method. The difference between the measured electrical resistances for photolithography processed ITO conductive tracks and those prepared by the present method lies within the range of measuring error.

A continuous UV-roll imprint lithography process was developed to fabricate ITO conductive patterns. The large-area metallic-roll stamp was fabricated by photolithography and electroforming, and was coated with an anti-adhesion layer. Using the present method, conductive ITO patterns were successfully generated and feasibility was confirmed by measuring the electrical resistance. The experimental results show that the present method can be applied for mass production of flexible electronics and flexible displays in large scale at low cost. Optimization of the process to obtain complicated patterns in nanoscale is a subject of ongoing research.

6.9 CONCLUSION

In this chapter, a UV-imprinting system for controlling the processing parameters was designed and constructed, and the effects of the processing conditions on the degree of photopolymerization, the elimination of microair bubbles, replication quality, and residual layer thickness were examined.

(1) A UV-imprinting system that consisted of a UV exposure unit, a mold and substrate jig unit with precision mechanical stage, a microscope and monitoring unit, and a system for applying pressure from the bottom was designed and constructed.

(2) The mechanism of photopolymerization was explained and the effect of varying the exposing dose on the curing state of the photopolymer was analyzed with FTIR spectroscopy. An exposure dose of $180\,mJ/cm^2$ is selected as the sufficient dose for the complete polymerization of the photopolymer.

(3) The microair bubble was eliminated using the preheating method. We heated the photopolymer to a temperature around $40°C$, at which the viscosity of photopolymer decreases and the trapped air bubble easily removed.

The effects of processing conditions on the replication quality and the uniformity of residual layer thickness of imprinted parts were analyzed. Finally, practical applications were presented.

REFERENCES

1. M. Wei, I. Su, Y. Chen, M. Chang, H. Lim, and T. Wu (2006) The influence of a microlens array on planar organic light-emitting devices. *Journal of Micromechanics and Microengineering*, 16, 368–374.

2. S. Kwon, X. Yan, A. M. Contreras, J. A. Liddle, G. A. Somorjai, and J. Bokor (2005) Fabrication of metallic nanodots in large-area arrays by mold-to-mold cross imprinting (MTMCI). *Nano Letters*, 5, 2557–2562.

3. N. Lee, S. Choi, and S. Kang (2006) Self-assembled monolayer as an antiadhesion layer on a nickel nanostamper in the nanoreplication process for optoelectronic applications. *Applied Physics Letters*, 88, 073101.

4. S. Moon, N. Lee, and S. Kang (2002) Fabrication of a microlens array using microcompression molding with an electroformed mold insert. *Journal of Micromechanics and Microengineering*, 13, 98–103.

5. S. Moon and S. Kang (2002) Fabrication of polymeric microlens of hemispherical shape using micromolding. *Optical Engineering*, 41, 2267–2270.

6. Y. Kono, A. Sekiguchi, Y. Hirai, S. Arasaki, and K. Hattori (2005) Study on nanoimprint lithography by the pre-exposure process (PEP). *Proceedings of SPIE* 5753, 912–925.

7. J.V. Crivello and K. Dietliker (1998) *Photoinitiators for Free Radical Cationic and Anionic Photopolymerisation*, Wiley, 64–70.

8. J. Lim (2006) Fabrication of UV-transparent mold of microlens array for image sensor. *Master Thesis, Yonsei University*.

9. T. Scherzer (2002) Real-time FTIR-ATR spectroscopy of photopolymerization reactions. *Macromolecular Symposia* 184, 79–98.

10. C. E. Corcione, A. Greco, and A. Maffezzoli (2004) Photopolymerization kinetics of an epoxy-based resin for stereolithography. *Journal of Applied Polymer Science* 92, 3484–3491.

11. S. Ahn, J. Cha, H. Myung, S. Kim, and S. Kang (2006). Continuous ultraviolet roll nanoimprinting process for replicating large-scale nano- and micro-patterns. *Applied Physics Letters* 89, 213101.

12. J. Lim, K. Jeong, S. Kim, J. Han, J. Yoo, N. Park, and S. Kang (2006) Design and fabrication of a diffractive optical element for objective lens of small form factor optical data storage device. *Journal of Micromechanics and Microengineering* 16, 77–82.

13. H. Kim, J. Lee, J. Lim, S. Kim, and S. Kang (2006) Design of microlens illuminated aperture array fabricated by aligned ultraviolet imprinting process for optical read only memory card system. *Applied Physics Letters* 88, 241114.

14. S. Kim, H. Kim, J. Lim, S. Kang, Y. Kim, R. Hendriks, A. Kastelijn, and C. Busch (2006) Elimination of jitter in microlens illuminated optical probe array using a filtering layer for the optical read only memory card system. *Japanese Journal of Applied Physics* 45, 1162–1166.

15. P. Nussbaum, I. Philipoussis, A. Husser, and H. P. Herzig (1998) Simple technique for replication of micro-optical elements. *Optical Engineering* 37, 1804–1808.

16. N. Lee, Y. Kim, and S. Kang (2004) Temperature dependence of anti-adhesion between the stamper with sub-micron patterns and the polymer in nanomoulding processes. *Journal of Physics D: Applied Physics* 37, 1624–1629.

17. P. Dannberg, G. Mann, L. Wagner, and A. Brauer (2000) Polymer UV-moulding for micro-optical systems and O/E-integration. *Proceedings of SPIE* 4179, 137–145.

18. N. Dumbravescu and M. Ilie (1999) Replication of diffractive gratings using embossing into UV-cured photo-polymers. *Proceedings of SPIE* 3879, 206–213.

19. St. Rudschuck, D. Hirsch, K. Zimmer, K. Otte, A. Braun, R. Mehnert, and F. Bigl (2000) Replication of 3D-micro- and nanostructures using different UV-curable polymers. *Microelectronic Engineering* 53, 557–560.

20. S. Kim and S. Kang (2003) Replication qualities and optical properties of UV-moulded microlens arrays. *Journal of Applied Physics D* 36(20), 2451–2456.

21. S. Kim, D. Kim, and S. Kang (2003) Replication of micro-optical components by ultraviolet-molding process. *Journal of Microlithography, Microfabrication, and Microsystems*, 2, 356–359.

22. S. Kim, H. Kim, and S. Kang (2006) Development of a UV-imprinting process for integrating a microlens array on an image sensor. *Optics Letters*, 31, 2710–2712.

23. M.C. Hutley, R. Hunt, R. F. Stevens, and P. Savander (1994) The moiré magnifier. *Pure and Applied Optics*, 3, 133–142.

24. N. Li, W. Wu and S.Y. Chou (2006) Sub-20 nm alignment in nanoimprint lithography using moiré fringe. *Nano Letters*, 6, 2626–2629.

25. H. Mifune, Y. Satoh, Y. Kiyosawa, and S. Satoh (2002) Fabrication of a high NA microlens with two substrates. *Microoptics News*, 19(1), 49–54.

26. I. Tanaka, Y. Iwasaki, M. Ogusu, K. Tamamori, Y. Sekine, T. Matsumoto, H. Maehara, and R. Hirose (1999) High-precision binary optical element fabricated by novel self aligned process. *Japanese Journal of Applied Physics* 38(12B), 6976–6980.

27. Y. Unno, Y. Sekine, E. Murakami, M. Ohta, and R. Hirose (1999) Application of large-scale binary optical elements to high-resolution projection optics used for microlithography. *Japanese Journal of Applied Physics*, 38(12B), 6968–6975.

28. Y. Sano, T. Nomura, H. Aoki, S. Terakawa, H. Kodama, T. Aoki, and Y. Hiroshima (1990) Submicron spaced lens array process technology for a high photosensitivity CCD image sensor. *Electron Devices Meeting, Technical Digest, International*, 283–286.

29. Z. D. Popovic, R. A. Sprague, and G. A. Neville Connell (1988) Technique for monolithic fabrication of microlens arrays. *Applied Optics*, 27, 1281–1297.

30. E. Yablonovitch, T. Gmitter, J. P. Harbison, and R. Bhat (1987) Extreme selectivity in the lift-off of epitaxial GaAs films. *Applied Physics Letters*, 51, 2222–2224.

31. H. Choe and S. Kim (2004) Effects of the n+ etching process in TFT-LCD fabrication for Mo/Al/Mo data lines. *Semiconductor Science and Technology* 19, 839–845.

32. M. Pudas, J. Hagberg, and S. Leppavuori (2004) Printing parameters and ink components affecting ultrafine-line gravure-offset printing for electronics applications. *Journal of European Ceramic Society* 24, 2943–2950.

33. D. Kim, S. Jeong, S. Lee, B. K. Kim, and J. Moon (2007) Organic thin-film transistor using silver electrodes by the inkjet printing technology. *Thin Solid Films* 515, 7692–7696.

34. J. Lee, S. Park, K. Choi, and G. Kim (2008) Nanoscale patterning using the roll typed UV-nanoimprint lithography tool. *Microelectronic Engineering*, 85, 861–865.

35. H. Ahn and L. J. Guo (2008) High-speed roll-to-roll nanoimprint lithography on flexible plastic substrates. *Advanced Materials* 20, 2044–2049.

36. H. Tan, A. Gilbertson, and S. Y. Chou (1998) Roller nanoimprint lithography. *Journal of Vacuum Science and Technology B*, 16, 3926–3928.

37. N. Lee, Y. Kim, S. Kang, and J. Hong (2004) Fabrication of metallic nanostamper and replication of nanopatterned substrate for patterned media. *Nanotechnology* 15(8), 901–906.

38. J. Han, S. Choi, J. Lim, B. S. Lee and S. Kang (2009) Fabrication of transparent conductive tracks and patterns on flexible substrate using a continuous UV-roll imprinting lithography. *Journal of Applied Physics D* 42(11), 115503.

39. Y. Loo, R. L. Willett, K. W. Baldwin, and J. A. Rogers (2002) Additive, nanoscale patterning of metal films with a stamp and a surface chemistry mediated transfer process: applications in plastic electronics. *Applied Physics Letters* 81, 562–564.

40. A. Kumar and G. M. Whitesides (1993) Features of gold having micrometer to centimeter dimensions can be formed through a combination of stamping with an elastomeric stamp and an alkanethiol "ink" followed by chemical etching. *Applied Physics Letters* 63, 2002–2004.

CHAPTER 7

High-Temperature Micro/ Nano Replication Process

Conventional replication processes typically utilize a polymer as the replication material. However, research on direct replication processes involving various electrical and optical materials has been given priority. These procedures preserve the characteristics of the materials while manufacturing the final products and offer relatively low cost and time consumption. In this chapter, high-temperature micro/nano replication processes will be considered using practical examples of metal nanopowder replication and glass molding.

7.1 FABRICATION OF METAL CONDUCTIVE TRACKS USING DIRECT IMPRINTING OF METAL NANOPOWDER

7.1.1 Introduction

Fabrication of conductive tracks is a key process in the manufacture of most electronic modules. In conventional fabrication processes, conductive tracks are fabricated by photolithography, which involves deposition of a conductive layer, substrate cleaning, photoresist (PR) coating, PR baking, PR exposure, conductive track etching, and PR stripping. Conventional conductive track fabrication is very complex and expensive, especially when applied to large areas such as display applications. Figure 7.1 shows the process flow of the conventional conductive track fabrication method. To overcome these limitations, various direct patterning processes such as inkjet patterning [1–4], nanoimprinting lithography (NIL) [5–8], and direct imprinting and sintering using a metal nanopowder have been proposed. In the inkjet process, a metal ink composed of metallic nanoparticles and solvent is directly imprinted on the substrate using an inkjet nozzle and a stage system, followed by a thermal treatment to evaporate the solvent and sinter the metal nanoparticles. Inkjet imprinting and sintering provide a

Micro/Nano Replication: Processes and Applications, First Edition. Shinill Kang.
© 2012 John Wiley & Sons, Inc. Published 2012 by John Wiley & Sons, Inc.

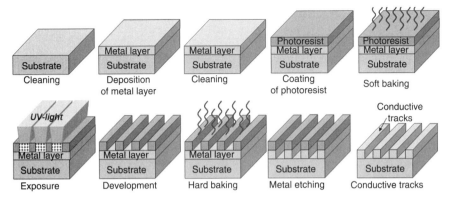

FIGURE 7.1 Process flow of conventional conductive track fabrication method using photolithography process.

simple fabrication process for conductive tracks, but the technique is limited by long fabrication times and the difficulty of fabricating ultrafine-line patterns. In the NIL process, a polymer material is patterned on the substrate by the physical contact and pressure of the stamp, which carries a negative image of the final pattern. NIL merges complex patterning procedures into a single imprinting process, which significantly simplifies the overall track patterning process, although metal deposition and etching are still involved. An imprinting and sintering method has been proposed, in which metal nanopowders are directly imprinted with a stamp that contains information on the tracks. This proposed process is simpler than conventional lithography and NIL-based track patterning. Although wet etching is still required for the removal of the unwanted residual layer, the proposed process does not require an etch barrier for metal etching or a sacrificial layer for the lift-off process. Also, the process significantly reduces the patterning time of metal nano-powders on the substrate compared with the inkjet process. The difference between the metal nanopowder inkjet technique and the imprinting and sintering technique lies not only in their effectiveness, but also in the micro-structure of the resulting conductive tracks. Because the proposed process provides a much denser microstructure, it significantly reduces the electrical resistivity. However, the process requires optimization, and possible defects should be analyzed and removed prior to industrial use.

In this study, an imprinting and sintering process for fabricating conductive tracks using a metal nanopowder was designed and analyzed, and the possible defects were considered. In addition, the effect of process conditions on the internal microstructures was examined, and the fabri-cated conductive tracks were evaluated to test the feasibility of the approach.

7.1.2 Direct Patterning Method Using Imprinting and Sintering

In the imprinting and sintering method, the metal nanopowder slurry is imprinted directly onto a large area substrate, which reduces the production cost and cycle time, improves the dimensional accuracy, and reduces the electrical resistivity.

Our new method consists of several steps. First, the powder slurry, which is a mixture of metal nanopowder and the solvent, is placed and spread on the glass substrate using a precision blade (Figure 7.2a). Next, the slurry is pressed using a silicon mold while heat is applied to sinter the metal nanopowders (Figure 7.2b). During the compression process, the excessive coated material is leaked out from the sandwich structure of substrate, metal nanopowder slurry, and mold and one can obtain uniform residual layer. Finally, the mold is removed, and conductive tracks are formed by removing the residual layer (Figure 7.2c–e). In this study, a metal nanopowder slurry composed of 67–74 wt% silver, 1.6–1.9 wt% copper, 21–26 wt% carbitol acetate, and 3.2–5.8 wt% Dowanol TPM was used as the imprinting material. The viscosity of the metal nanopowder slurry was 1270 cp in room temperature. An O_2 plasma treatment for increasing adhesion properties between glass substrate and imprinted metal conductive track was performed before the imprinting process. The thickness of residual layer was 800 ± 30 nm after the imprinting process and a nitric-acid-based etchant (CR4, Cyantek, U.S.A) was used for the residual layer removal process. The etching rate was 100 nm/s and the etching time was $\sim$8.5 s.

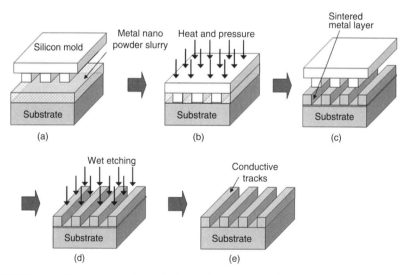

FIGURE 7.2 Process flow of imprinting and sintering method using nanometal powder: (a) coating of metal nanopowder on a substrate; (b) heating and pressing process; (c) demolding process; (d) removing residual layer using wet etching process; (e) fabricated conductive tracks. Reprinted with permission from Ref. [9]. Copyright 2009, Institute of Physics.

7.1.3 Imprinting and Sintering System

We designed and fabricated inverted conductive pattern structures for the silicon mold. To enable measurement of resistance of the conductive tracks after the fabrication process, each conductive track had square-shaped pad structures at both ends. The fabricated silicon mold (Figure 7.3a) had a

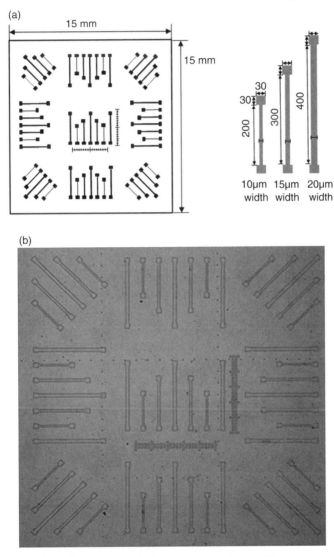

FIGURE 7.3 Schematic diagram of designed mold: (a) shape of line patterns in a unit cell; (b) microscope image of the unit cell of the fabricated silicon mold; (c) microscope image of the line patterns in the silicon mold. (d) SEM image of the line patterns in the silicon mold. Reprinted with permission from Ref. [11]. Copyright 2007, American Institute of Physics. Reprinted with permission from Ref. [9]. Copyright 2009, Institute of Physics.

(c)

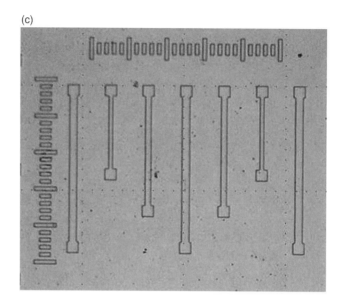

(d)

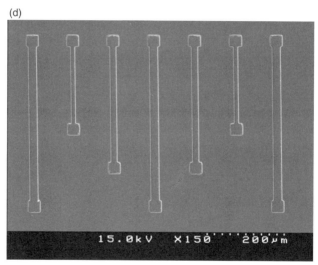

FIGURE 7.3   (*Continued*)

conductive track thickness of 400 nm and line widths of 10, 15, and 20 μm. The pads were square-shaped with dimensions of 30 × 30 μm. The mold was fabricated on a four-inch silicon wafer consisting of 16 identical 15 × 15 mm cell structures. Each cell was diced to obtain the final silicon mold. Figure 2b shows a microscopic image of the final mold, which was fabricated using photolithography and reactive ion etching (RIE) [10].

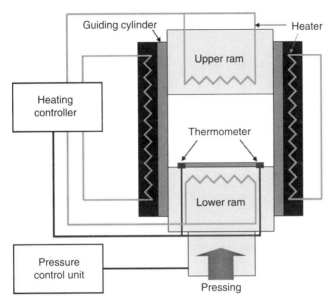

FIGURE 7.4   Schematic diagram of test setup. Reprinted with permission from Ref. [9]. Copyright 2009, Institute of Physics.

A system for imprinting and sintering of conductive tracks that could apply pressure and heat simultaneously during direct metal patterning was developed. Figure 7.4 shows a schematic of the apparatus setup [12,13]. The apparatus consisted of a band heater, a guiding cylinder for the various jigs, a pressure cylinder, a lower ram to transfer pressure, a thermometer, and finally a pressure and heating control unit. The actual testing temperature was measured and controlled by measuring the temperature of the upper ram. The hydraulic pump was the pressure source; during the fabrication process, the pressure was monitored and controlled accordingly.

To increase the releasing properties of the silicon mold, a self-assembled monolayer (SAM) of (tridecafluoro-1,1,2,2-tetrahydrooctyl)trichlorosilane was coated onto the mold surface as an anti-adhesion layer [14]. The enhancement of the releasing properties by the addition of the SAM-coated layer was evaluated by measuring the contact angle using a previously reported method [15]. The contact angles were found to be 70.37° for the bare silicon mold and 115.86° for the SAM-coated silicon mold, which confirms the effectiveness of the SAM-coating process. However, the processing temperature used for metal nanopowder imprinting is higher than those used in other polymer imprinting processes, and may degrade the releasing properties of the SAM layer. To test the thermal stability of the SAM layer as a function of temperature, we measured the contact angles of deionized water on a SAM-coated silicon mold after thermal treatment at various

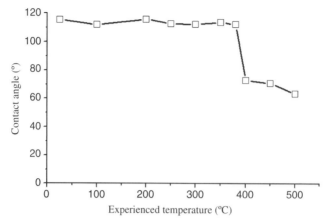

FIGURE 7.5 Contact angle of deionized water on a thermally treated SAM-coated silicon mold as a function of the temperature used for thermal treatment; the contact angle of bare silicon was 70.37°. Reprinted with permission from Ref. [11]. Copyright 2007, American Institute of Physics.

temperatures. Figure 7.5 shows the change of the contact angle of the SAM layer as a function of temperature. The contact angle of the SAM decreased markedly at temperature above 380°C, indicating deterioration of the SAM material in this temperature regime, which suggested that the processing temperature should be below 380°C.

Figure 7.6 shows the comparison of SEM images of imprinted tracks using (a) bare silicon mold and (b) silicon mold coated with anti-adhesion layer. A missing pattern and incomplete replication due to the sticking of mold were shown in Figure 7.6a, but the replication quality was improved markedly as

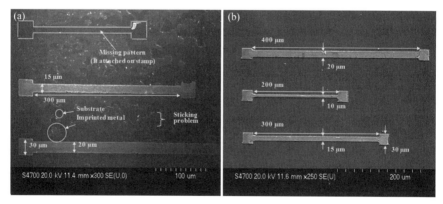

FIGURE 7.6 Comparison of SEM images of nanoimprinted metallic conductive tracks using (a) bare silicon mold and (b) silicon mold coated with anti-adhesion layer (SAM of (tridecafluoro-1,1,2,2-tetrahydrooctyl)trichlorosilane). Reprinted with permission from Ref. [11]. Copyright 2007, American Institute of Physics.

shown in Figure 7.6b. These results clearly show the effectiveness of SAM as an anti-adhesion layer for this method.

7.1.4 Defect Analysis and Process Design

The silver–copper alloy metal nanopowder was mixed with solvent to form a slurry [16]. Therefore, solvent removal was taken into account during this experiment. Gas pores may be generated by evaporated solvents that escape between the glass substrate and the silicon mold. In addition, solvent may become trapped within the imprinted sample and leave defect trails of random microchannels.

Figure 7.7 shows SEM images of a metal nanopowder layer that was imprinted without considering the solvent removal process. We observed trails of the solvent's escape from the sintered metal nanopowder, resulting in the formation of random microchannels.

To prevent microchannel formation within the imprinted conductive tracks, we optimized the pattern formation by adjusting temperature and pressure. Figure 7.8 shows the history of the actual temperature and applied pressure conditions, and indicates clearly that pressure was applied after the temperature had saturated up to the sintering temperature. During the temperature saturation period, no pressure was applied and volatile solvents evaporated, preventing the generation of defect gas-pore microchannels. Another reason for this period was to eliminate unsteady heating before the saturated temperature was reached, as well as to prevent damage to the substrate and silicon mold due to thermal stress.

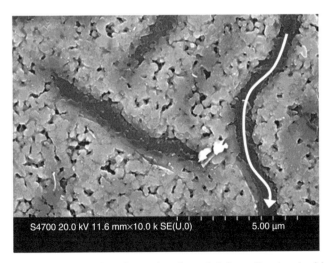

S4700 20.0 kV 11.6 mm×10.0 k SE(U,0)   5.00 µm

FIGURE 7.7 SEM images of random microchannel defects. Reprinted with permission from Ref. [9]. Copyright 2009, Institute of Physics.

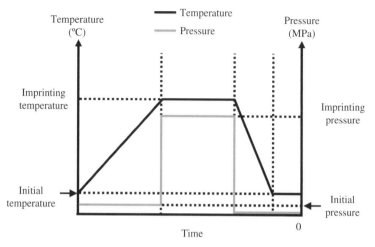

FIGURE 7.8 Histories of process conditions. Reprinted with permission from Ref. [9]. Copyright 2009, Institute of Physics.

7.1.5 Analysis of Imprinted Conductive Tracks

(1) Geometrical properties

The applied pressure influenced the pattern fidelity. Figure 7.9 shows the surface profile of the reversed silicon mold, a conductive track with a line width of $20\,\mu m$ fabricated with applied pressures of 34 and $150\,MPa$. The silicon mold had a pattern depth of $400\,nm$, but the pattern fabricated with an applied pressure of $34\,MPa$ had a height of $345\,nm$, which deviated by 15% from the original silicon mold. The pattern height of the conductive track fabricated with an applied pressure of $150\,MPa$ was $398\,nm$, amounting to a deviation of 1%.

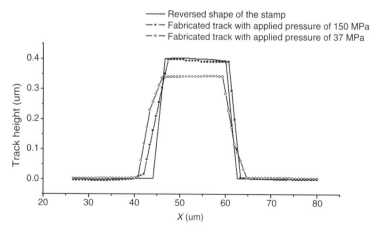

FIGURE 7.9 Measurement results of fabricated conductive tracks using white light interferometer. Reprinted with permission from Ref. [9]. Copyright 2009, Institute of Physics.

(2) **Microstructure and electrical resistivity**

To analyze the effect of pressure on the microstructures of the conductive tracks fabricated by imprinting and sintering, we varied the applied pressure to 37, 74, 112, and 150 MPa, with a temperature of 310°C and duration time of 1 h.

The microstructures of the fabricated conductive tracks at different pressures were analyzed. Figure 7.10 shows the differences in the sintered structures

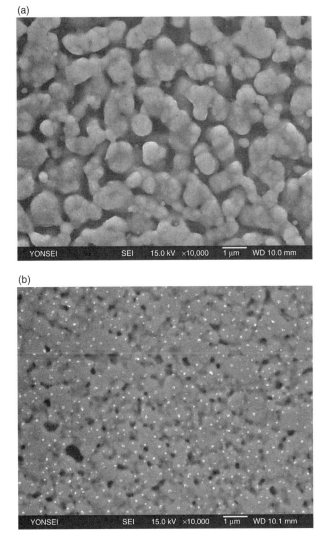

FIGURE 7.10   SEM images of surfaces of imprinted conductive tracks with different applying pressure: (a) 37 MPa; (b) 74 MPa; (c) 112 MPa; (d) 150 MPa. Reprinted with permission from Ref. [9]. Copyright 2009, Institute of Physics.

(c)

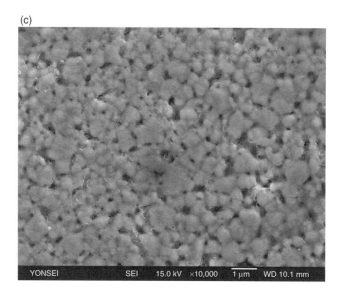

(d)

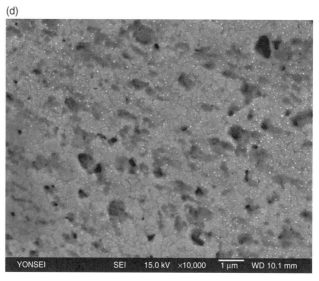

FIGURE 7.10   (*Continued*)

due to varying pressure with constant temperature and duration time. Figure 7.10a shows the weakest pressure applied during the sintering process in our experiment. This was the stage in which the metal nanopowders began to interact, with the relationships between the powders contact without any deformation of the powders. As shown in Figures 7.10a–d, an increase in the sintering pressure increased the contact area between the metal nanopowders. As shown in Figure 7.10d, when the applied pressure was 150 MPa, metal

FIGURE 7.11 SEM images of inkjet conductive tracks. Reprinted with permission from Ref. [9]. Copyright 2009, Institute of Physics.

nanopowders formed sintered structures with grain sizes ranging from 200–500 nm. The microstructure of the direct-patterned conductive tracks was much denser than that of the inkjet conductive tracks shown in Figure 7.11.

Figure 7.12a shows an SEM image of sample conductive tracks fabricated under a pressure of 150 MPa, which had the best sintered structure in our experiments. To analyze the intermolecular structure of conductive tracks sintered at 150 MPa, a focused ion beam (FIB) milling process was used to acquire cross-sectional SEM images. As shown in Figure 7.12b, sintered conductive tracks indicated that the pores formed had negligible size, with a continuous and dense structure.

Thus, the imprinting and sintering method enabled fabrication of conductive tracks with better geometrical properties and microstructure than those fabricated by a similar process without the application of pressure using the inkjet process.

Finally, Karl Suss Probe Station PM5 was used to measure the electrical conductive properties of track resistances. In addition, by analyzing the pattern structure, we calculated the electrical resistivity. Figure 7.13 shows the probe station used for our analysis. We measured conductivity at the multiple points (4–6 points) of the five different samples that were fabricated at the same process conditions.

Figure 7.14a shows the effects of imprinting temperature on the electric resistivity for fixed pressure and processing time. It is clear that the resistivity decreases linearly with increases in the imprinting temperature. The mean electric resistivity was found to decrease from 258–42.71 $\mu\Omega \cdot cm$ with increases in the imprinting temperature from 250–310°C at a sintering pressure of 37 MPa and an imprinting time of 1 h. This means that increasing the process temperature results in increase of degree of sintering and reduced porosity. However, the use of high temperatures can damage other preprocessed structures in the electronic circuit. Figure 7.14b shows the effects of imprinting

(a)

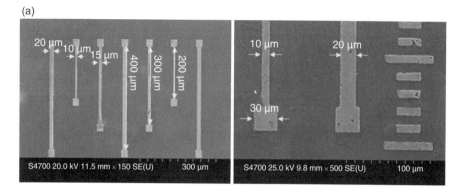

(b)

FIGURE 7.12 (a) SEM images of fabricated conductive tracks with pressure of 150 MPa, temperature of 310°C (b) cross-sectional view of imprinted and sintered conductive track. Reprinted with permission from Ref. [9]. Copyright 2009, Institute of Physics.

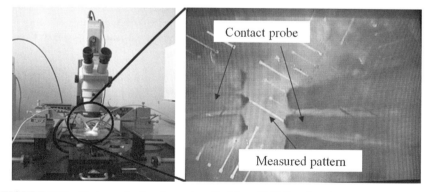

FIGURE 7.13 Photograph of probe station; Karl Suss PM5. Reprinted with permission from Ref. [9]. Copyright 2009, Institute of Physics.

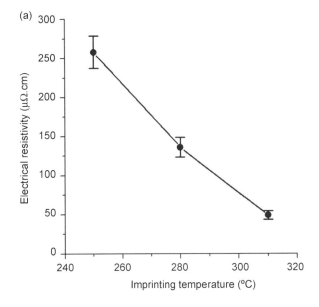

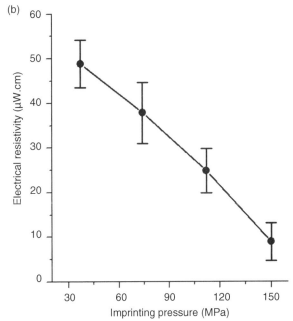

FIGURE 7.14 Effects on the electric resistivity of varying (a) the imprinting temperature for a pressure of 37 MPa and an imprinting time of 1 h; (b) the applied pressure for a temperature of 310°C and an imprinting time of 1 h. Reprinted with permission from Ref. [11]. Copyright 2007, American Institute of Physics.

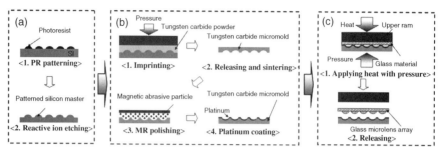

FIGURE 7.15 Schematic diagram of the proposed microthermal forming process, which consists of: (a) original master fabrication; (b) WC micromold fabrication using imprinting and sintering process; (c) glass microlens array fabrication by microthermal forming process. Reprinted with permission from Ref. [17]. Copyright 2008, World Scientific.

pressure on the electric resistivity for fixed temperature and processing time. The resistivity was found to decrease from $42.71–8.95\,\mu\Omega\cdot\text{cm}$ with the increase in the applied pressure from 37–150 MPa at an imprinting temperature of 310°C and an imprinting time of 1 h.

7.1.6 Conclusions

We evaluated the potential of the imprinting and sintering method using metal nanopowder process to overcome the limitations of conventional conductive track fabrication methods. The effect of the process conditions on pattern formation was investigated, and the feasibility of the industrial process was confirmed. The analysis was based on the sintering and pattern fidelity of the conductive tracks fabricated under different process conditions. Conductive tracks fabricated under an applied pressure of 150 MPa had the most densely sintered structure, with a pattern height deviation of 1% from the original silicon mold. The limitation of the present process is that the selected processing values of the temperature and pressure are rather high for use in conventional circuit fabrication processes. These processing values could be decreased by selecting the metal nanopowder with smaller particle size because the sintering conditions were affected by the material properties and the processing conditions [10]. In addition, silicon mold, which was used for this study, is relatively fragile for mass production. To overcome this limitation, a metallic mold for the presented method using various surface treatment solutions for anti-adhesion such as diamond-like coating (DLC) is under development. Although the processing conditions and the mold were not sufficient for some applications, we thought our experimental results were very useful for other research groups in the fields of electric circuit fabrication, nanoimprinting, metallic micro/nanostructure technology and so on. Further improvements such as increases in the patterning area, steadiness of the fabrication process, and reduction in the temperature and pressure required, controlling the uniformity of residual layer thickness and etching process are subjects of ongoing research. The whole process of microthermal forming are schematically illustrated in Fig 7.15

7.2 GLASS MOLDING OF MICROLENS ARRAY

7.2.1 Introduction

Remarkable progress has recently been made in the technology of polymer microlens arrays, which are used in optical data storage, digital displays, and optical communications. However, the use of glass in optical components offers many advantages over the use of polymers. For instance, the optical properties of glass materials, such as the refractive index and the thermal expansion coefficient, are more robust with respect to temperature change. Furthermore, glass materials have excellent acid and moisture resistance. Accordingly, the demand for glass microlens arrays has been increasing in applications that require high optical performance and environmental stability, such as high-density optical data storage and multimedia projector components. In the past, glass microlens arrays were mainly produced by reactive ion etching (RIE) [19] or diamond turning [20]. However, mass production at low cost has proved to be rather difficult due to the limited etching range of RIE processing parameters and the low production rate of the diamond turning process.

Microthermal forming is a promising replication process for the mass production of high-quality glass microlens arrays at relatively low cost [21]. A tungsten carbide (WC) micromold is commonly used in this process [22], and direct machining is generally utilized for WC micromold fabrication [23].

Direct machining is not appropriate for the fabrication of microlens cavities with diameters of $100\,\mu m$ or less. To overcome this limitation, in this study, a WC micromold was fabricated by the imprinting and sintering process [24]. Glass microlens arrays with lens diameter under $100\,\mu m$ were then replicated by microthermal forming under various processing conditions. The effects of microthermal forming process conditions on the form accuracy and surface roughness of glass microlens arrays were examined and analyzed. Finally, to verify the effectiveness of the proposed process, a glass microlens array for laser beam focusing was designed and fabricated, and a focused laser beam profile using the fabricated microlens array was measured and analyzed.

7.2.2   Fabrication of Master Patterns

(1) Lens master fabrication by RIE method

The proposed pressure-forming process involves the replication of a prepared master surface shape with a mixture of tungsten carbide powder and a binding agent. In this step of the process, the master material must be sufficiently hard to prevent damage to the master pattern by tungsten carbide powder. At the same time, the master pattern must be easy to fabricate. In this study, silicon was selected as the master material because it satisfies both requirements.

PR patterns for the microlens arrays were manufactured via the reflow method [23–26]. A propylene glycol monomethyl ether acetate (PGMEA)-based positive photoresist was used, and the resist was first baked on a hot plate at 170°C for 60 s and then in an oven at 170°C for 30 min. RIE [27,28] was then utilized to fabricate the microlens array masters on silicon wafers. Figure 7.16 illustrates the RIE process. To produce a lens with the desired geometrical properties, such as diameter, height, and radius of curvature, the diameter and height of the initial PR pattern as well as the choice of process conditions are important.

Figure 7.17 shows scanning electron microscopy images (SEM) of PR patterns with lens diameters of 36 and 96 μm produced by the reflow method. The fabrication of the microlens array masters on the silicon wafers was dependent on the transfer of the reflowed microlens arrays by RIE. Before initiating the RIE process, an additional heat treatment procedure was applied to the PR patterns. Without the prior hardening of the PR, the RIE process yielded silicon patterns with poor surface quality, as shown in Figures 7.18a and b. The hardened PR produced a better surface finish on the silicon patterns, as shown in Figures 7.18c and d.

(2) Etching selectivity during RIE process

To control the shape of the etched lens, it is necessary to maintain the selectivity ratio of the silicon and PR etching processes at 1:1. In the RIE process, SF_6 gas determines the silicon etch rate, and O_2 gas determines

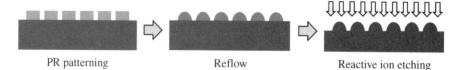

PR patterning   Reflow   Reactive ion etching

FIGURE 7.16   Schematic diagram of the fabrication of a reactive-ion-etched lens.

(a)

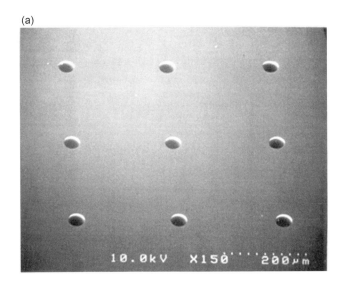

(b)

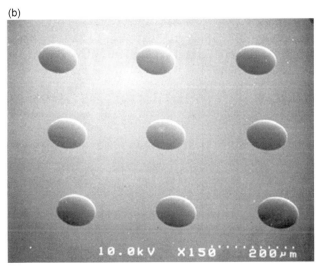

FIGURE 7.17   SEM images of the mother lens array: (a) lens diameter of 36 μm and (b) lens diameter of 96 μm.

the PR etch rate. Although gas pressure and power also affect the etch rates, the influence of the gas ratio on selectivity is predominant.

Selectivity can therefore be controlled by the ratio of the SF_6 and O_2 gases. The SiO_2 layer on the silicon wafer, formed by oxidation during the reflow process, has an adverse effect on RIE results. As the etch rate of the SiO_2 layer is 10 times slower than that of the silicon, selectivity control is difficult to accomplish with the silicon oxide layer. Therefore, the SiO_2 layer must be removed by HF etching before initiating the RIE process. Figure 7.19 compares the surface profiles of the PR following the reflow process and the lens master produced by the RIE process when a selectivity ratio of 1:1 is achieved.

(a)

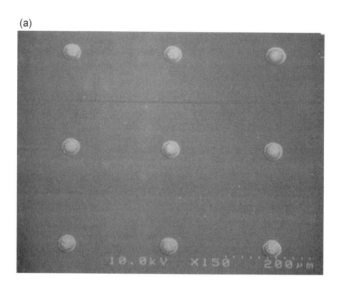

(b)

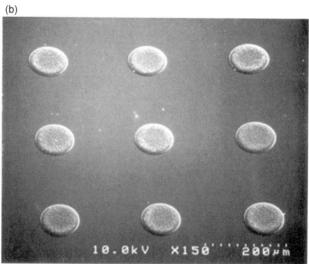

FIGURE 7.18 SEM images of the reactive-ion-etched lens array: (a) and (c) lens diameter of 36 μm; (b) and (d) lens diameter of 96 μm; (a) and (b) reflow condition is 170°C, 60 s on hot plate; (c) and (d) reflow condition is 170°C, 60 s on hot plate and 170°C, 30 min in oven.

(c)

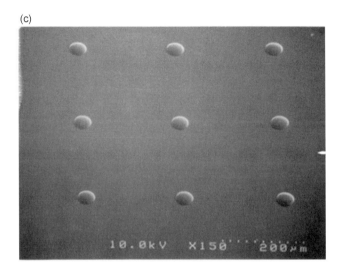

(d)

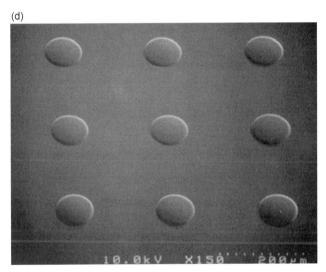

FIGURE 7.18 (*Continued*)

7.2.3 Fabrication of Tungsten Carbide Core for Microglass Molding

7.2.3.1 Fabrication Process of Tungsten Carbide Core Glass

molding press processes are normally conducted at temperatures between 300 and 1400°C. For the sake of precision, when glass lenses are fabricated by this method, the die material must have sufficient strength, hardness, and accuracy at high temperatures and pressures. Good oxidation resistance, low thermal expansion, and high thermal conductivity are also required. Tungsten

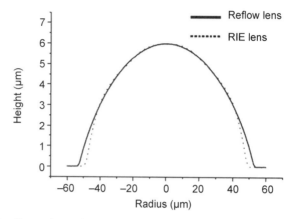

FIGURE 7.19 Comparison of the surface profile of the reflow lens and the reactive-ion-etched lens for a selectivity of 1:1.

carbide is a suitable die material because it possesses all of these properties [29,30].

The tungsten carbide cores are produced by powder metallurgy, most commonly with the addition of a metallic binder (usually cobalt) during liquid-phase sintering. In this process, the metallic binder improves the toughness of the cores as well as the sintering quality in the liquid phase to fully dense material. A tungsten carbide core was fabricated by the following procedure [31]. Tungsten carbide powders with three different grain sizes (0.2, 0.5, and 0.8 µm) were mixed with ready-to-press (RTP) powder and bonding agent. The powder mixture was then pressure-formed using the silicon mold master.

Compression pressure was applied to the die as shown in Figure 7.20. The forming pressure must be high enough to assure the replication quality and provide the required bonding force between adjacent tungsten carbide grains

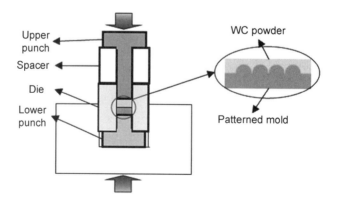

FIGURE 7.20 Schematic diagram of the pressure forming process.

and bonding agents. However, if the forming pressure is excessively high, defects in the silicon master may occur. A sintering process follows pressure forming. Sintering is a process whereby compressed metal powder is heated in a controlled-atmosphere furnace to a temperature below its melting point, but sufficiently high to allow bonding. In this study, a vacuum was maintained in the chamber to prevent oxidation. Presintering was carried out at 350°C for 10 h to remove the bonding agents and promote dehydration. The temperature was then slowly raised, and the sintering was performed at 1400°C for 50 min. Figure 7.21 shows SEM images of the resulting tungsten carbide lens arrays with diverse diameters.

(a)

(b)

FIGURE 7.21   SEM images of the tungsten carbide lens array: (a) lens diameter of 58 μm; (b) lens diameter of 96 μm; (c) lens diameter of 182 μm.

(c)

FIGURE 7.21   (*Continued*)

7.2.3.2 Measurement of Shrinkage After Sintering Process   In order to evaluate the shrinkage of the tungsten carbide core due to the sintering process, the geometries of the silicon master and the fabricated tungsten carbide core were measured and compared. For test purposes, another type of tungsten carbide core (with a *V*-groove pattern) was fabricated along with the lens array pattern. Table 7.1 lists the measured dimensions, and Figure 7.22 shows SEM images of the *V*-groove silicon master and the fabricated tungsten carbide core. As Figure 7.22 indicates, the same amount of shrinkage occurred in both the pitch and the depth. The lens master and the fabricated core were also measured by a three-dimensional scanning optical profiler (Wyko NT2000).

To examine the shrinkage of the lens core, master lenses with varying diameters (from about 48–300 μm) were prepared. As can be inferred from Table 7.2, the lens cores underwent isotropic shrinkage for all the samples, regardless of the lens diameter. On the basis of these results, it is evident that the shrinkage rate of a tungsten carbide lens core during the sintering process is constant with respect to diameter and depth, and thus, the final shape of the lens core is identical to that of the master. Therefore, shape change during the sintering process can be compensated by proper silicon master design.

TABLE 7.1   Geometrical Properties of the *V*-Groove Silicon Master and WC Core

	Master	WC Core	Shrinkage (%)
Pitch (μm)	250.2	192.14	23
Depth (μm)	89.4	92.58	23

(a)

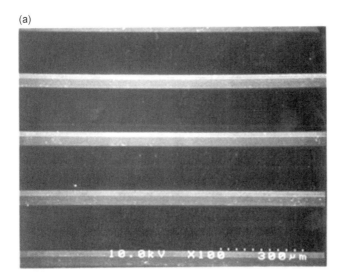

(b)

FIGURE 7.22   SEM images of the *V*-groove: (a) Si *V*-groove master; (b) tungsten carbide *V*-groove core.

7.2.4 Surface Finishing and Coating Process of Tungsten Carbide Core

A high-quality core surface is essential for producing a high-quality replicated lens array. Asperities on the core surface can cause a release problem for a replicated component and degrade the surface quality of the product. Especially during the liquid-phase sintering process, massive grain growth of the tungsten carbide powders is unavoidable, and this produces a

TABLE 7.2 Geometrical Properties of Lens Silicon Master and WC Core

	Sample	Master	WC Core	Shrinkage (%)
1	Diameter (μm)	300.0	224.6	25.1
	Height (μm)	150.0	113.2	24.5
2	Diameter (μm)	24.0	182.0	24.2
	Height (μm)	60.0	45.5	24.2
3	Diameter (μm)	180.0	137.0	23.9
	Height (μm)	30.0	22.9	23.7
4	Diameter (μm)	76.7	58.1	24.3
	Height (μm)	7.2	5.5	24.0
5	Diameter (μm)	57.5	43.4	24.5
	Height (μm)	7.2	5.4	24.2
6	Diameter (μm)	47.6	35.9	24.6
	Height (μm)	7.0	5.3	24.4

coarse-textured surface. Therefore, a surface-smoothing process is needed to improve the mold quality. However, it is difficult to smooth the surfaces of a three-dimensional microstructure, including the microlens arrays fabricated in this study. In particular, traditional finishing techniques such as manual lapping, grinding, and pad polishing are not appropriate for tungsten carbide cores for microlens arrays due to the small feature size and high hardness of the workpiece. Moreover, the finishing process must not alter the original shape of the workpiece.

In this study, forced dragging of fine abrasives across the inside surface of the core resulting in material removal is suggested. To accomplish this, a technique based on a magnetic abrasive finishing process [32] is introduced, in which material is removed in the presence of a magnetic field in the machining zone, with magnetic abrasive particles acting as the cutting tools. The magnetic abrasives are produced by bonding abrasives to a ferromagnetic material. In this case, commercial carbonyl iron powders (mean diameter = 7 μm) and diamond particles (mean diameter = 1 μm) are bonded together using cyanoacrylate glue in a weight ratio of 6:3:1. The bonded material is then milled to obtain the magnetic abrasives [33,34].

Figure 7.23 shows a schematic diagram of the applied finishing method. An electromagnet with a hollow cylindrical shape is connected to a current source. A long circular magnetic shaft is inserted into the cylinder and supported by two ball bearings at both ends. The shaft is connected coaxially to the spindle of a milling machine. The working tip of the shaft is located above the tungsten carbide core. In this configuration, a strong vertical magnetic field is generated from the tip to the core, and a polishing brush is formed from the polarized magnetic abrasives.

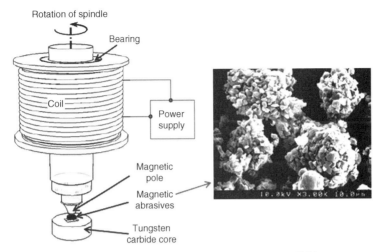

FIGURE 7.23   Schematic diagram of the magnetic-field-assisted polishing process and SEM image of the bonded magnetic abrasives.

In the experiments, the rotating shaft diameter was 5 mm at the tip. The measured magnetic flux density at the tip was about 0.45 T. The distance between the tip and the core was 0.4 mm. The spindle speed was 2000 rpm, and the machining time was 30 min.

7.2.5 Comparison of Surface Roughness Before and After Finishing Process

The surface roughness after sintering is affected by the grain size. To investigate the effect of grain size, tungsten carbide cores with three different grain sizes (0.2, 0.5, and 0.8 μm) were sintered. The surface roughnesses (*Ra*) of the cores produced with grain sizes of 0.2, 0.5, and 0.8 μm were 0.16–0.21 μm, 0.28–0.32 μm, and 0.50–0.55 μm, respectively. As expected, the best surface quality was achieved with the smallest grain. Figures 7.24a and b show SEM images of tungsten carbide lens arrays, each with a diameter of 58 μm, fabricated with RTP powders with 500 and 200 nm grain sizes, respectively. Although little improvement in surface quality is observed between the two figures, there is also a limit to which surface quality can be improved solely by decreasing the grain size. For further improvement in surface quality, the proposed finishing method was carried out on the core produced with the 500 nm grain RTP powder.

The surface roughness of the core before and after the finishing was measured with an SEM. Figures 7.25a and b show SEM images of the tungsten carbide lens array (with a diameter of 58 μm) fabricated with the 500 nm grain RTP powder, before and after finishing. As can be seen, the

(a)

(b)

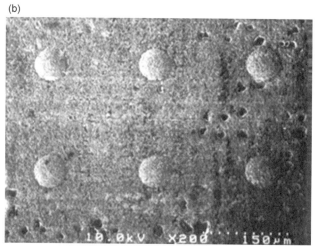

FIGURE 7.24   SEM images of the tungsten carbide lens array with a diameter of 58 μm made using RTP powder with the (a) grain size 500 nm and (b) grain size 200 nm.

surface was smoother after the finishing process. As the surface of the cavity could not be accessed directly by an atomic force microscopy device, precise evaluation of the surface roughness of the mold itself was carried out by measuring the surface roughness of polymer replicas of the tungsten carbide cores, which were fabricated via a UV-imprinting process. Figures 7.26a–d show AFM images of the polymer replicas of the tungsten carbide lens array. Figures 7.26a and b show the scanned images of the plane before and after finishing, respectively. The surface roughness decreased considerably after the finishing process; *Ra* and *Rmax* changed from 0.128 and

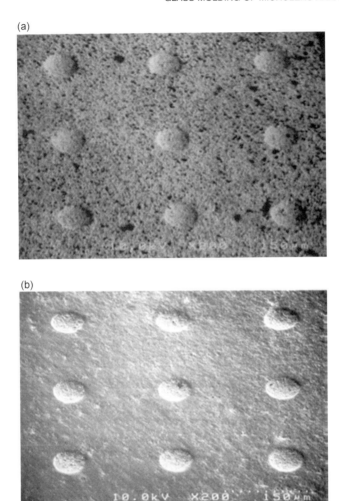

FIGURE 7.25 SEM images of the tungsten carbide lens array with a diameter of 58 μm made using RTP with a grain size of 500 nm: (a) before polishing; (b) after polishing.

2.1 μm before to 5.5 and 58.8 nm after finishing. Figures 7.26c and d show the flattened images of the cavities before and after finishing, and here, Ra and $Rmax$ changed from 0.144 and 1.839 μm before to 53.6 nm and 0.933 μm after. The surface roughness of the plane and the cavity, measured after the sintering process only, were similar. After finishing, however, the grain boundaries and defects were greatly diminished, and the surface roughness improved for both surfaces. The smoothness of the plane is better than that of the cavity region because of the slow material removal rate in the cavity.

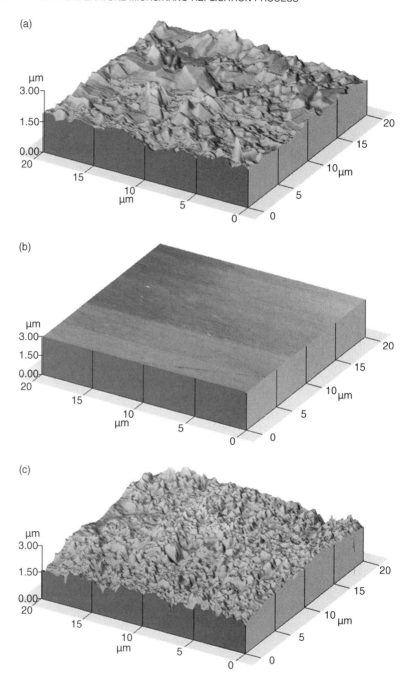

FIGURE 7.26   AFM images of tungsten carbide lens array made using RTP powder with a grain size of 500 nm: (a) and (b) plane part; (c) and (d) lens cavity part; (a) and (c) before polishing; (b) and (d) after polishing (areal surface mean/p–v roughness in (a), (b), (c) and (d): 0.128 μm/2.1 μm, 5.5 nm/58.8 nm, 0.144 μm/1.839 μm, 53.6 nm/0.933 μm).

(d)

FIGURE 7.26   (*Continued*)

As a final step, a surface coating procedure is performed to prevent melted glass from sticking to the core surface in glass molding applications [35–38]. In this study, the surface of the core was coated with platinum.

7.2.6 Fabrication of Glass Microlens Array by Microthermal Forming Process

Figure 7.27 shows a schematic diagram of the microthermal forming system used in this study. The machine consists of a system for applying pressure, a WC micromold, a jig for the WC micromold, infrared lamps for heating, a load cell, and a thermocouple. A glass material was chosen. The glass

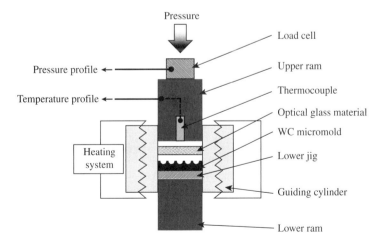

FIGURE 7.27   Schematic diagram of the microthermal forming machine. Reprinted with permission from Ref. [17]. Copyright 2008, World Scientific.

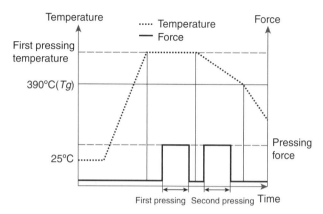

FIGURE 7.28 The variations of the temperature and compression force during the micro-thermal forming process. Reprinted with permission from Ref. [17]. Copyright 2008, World Scientific.

transition temperature (*Tg*) and yield temperature (*At*) of the glass (K-PSK100, Sumita Optical Glass, Inc.) are 390 and 415°C, respectively.

The graph in Figure 7.28 shows the temperature and the applied compression force during the microthermal forming process (total processing time: 12 min). The two steps are as follows: (1) pressing near the yield temperature with a precisely controlled force; and (2) pressing near the glass transition temperature. The second pressing is carried out while the material cools from the first pressing temperature to the glass transition temperature.

If the molding process is completed at a temperature close to the yield temperature, the problem of material shrinkage becomes critical, as thermal expansion of the glass is significant above the glass transition temperature. In our technique, the first and second pressing forces are the same.

The compression force had no significant effect on the replication quality for compression forces in the range 50–200 kgf at a first pressing temperature of 415°C. In our experiment, the compression force was fixed at 50 kgf.

Figure 7.29 shows three-dimensional surface profiles of glass microlenses fabricated at various first pressing temperatures with a 50 kgf compression force. The glass microlens fabricated at the first pressing temperature of 400°C (Figure 7.29a) was defective because insufficient glass material entered the cavity. At the first pressing temperature of 415°C (Figure 7.29b), the glass microlens with lowest surface roughness was produced. In this case, no surface defects occurred; i.e., no sticking occurred between the glass material and the WC micromold surface. At the first pressing temperatureof 430°C (Figure 7.29c), the surface roughness was excessive; a spire pattern caused by sticking between the glass material and the WC micromold surface appeared

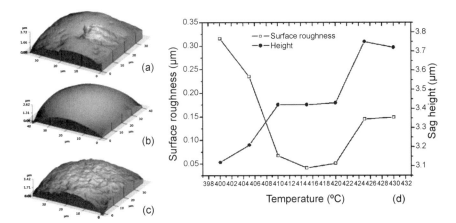

FIGURE 7.29 Effects of varying the first pressing temperature on the sag height and surface roughness of glass microlenses replicated with a compression force of 50 kgf: (a), (b), and (c) show the AFM images of the replicated microlenses' surfaces for various first pressing temperatures, and (d) shows the relationships between the first pressing temperature, the surface roughness (*Ra*) and sag height of the glass microlenses. Reprinted with permission from Ref. [17]. Copyright 2008, World Scientific.

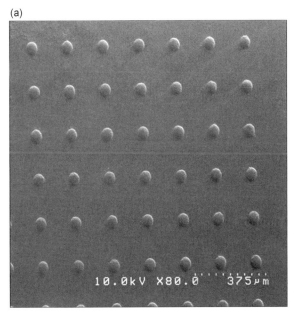

FIGURE 7.30 (a) SEM image of the formed glass microlens array with lens diameter of 58 μm and (b) surface profiles of the formed glass microlens (process condition: 415°C first pressing temperature, 50 kgf compression force) and WC micromold cavity. Reprinted with permission from Ref. [17]. Copyright 2008, World Scientific.

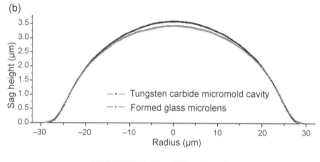

FIGURE 7.30 (*Continued*)

on the glass microlens surface. The graph in Figure 7.29d shows the variations of the sag height and surface roughness of the fabricated glass microlenses with respect to the first pressing temperature at a compression force of 50 kgf. The sag height increased as the first pressing temperature increased because the fluidity of the glass material increases. However, between 410 and 420°C

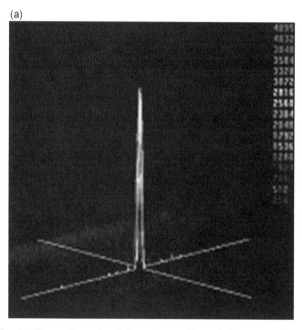

FIGURE 7.31 (a) Three-dimensional intensity profile and (b) comparison of diffraction limited beam intensity profile and normalized intensity profile of the focused laser beam (wavelength of 609 nm) using a replicated glass microlens with a diameter of 58 μm, a sag height of 3.42 μm, and a surface roughness (*Ra*) of 44 nm. Reprinted with permission from Ref. [17]. Copyright 2008, World Scientific.

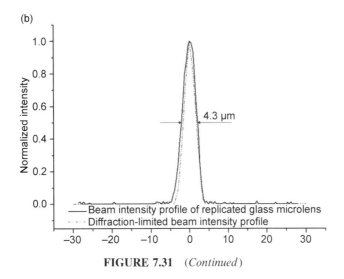

(b)

FIGURE 7.31   (*Continued*)

(which is close to the yield temperature of 415°C), the sag heights were similar. The glass microlens array with the best surface roughness was obtained at the yield temperature (415°C) of the glass material. Measurements taken from 10 glass microlens samples (process conditions: 415°C first pressing temperature, 50 kgf compression force) yielded an average sag height of 3.424 μm (standard deviation: 0.022 μm), an average radius of curvature (measured by a surface profiler) of 145 μm (standard deviation: 5.6 μm), and an average diameter of 58.04 μm (standard deviation: 0.19 μm). Figure 7.30 shows an SEM image of a replicated glass microlens array (process conditions: 415°C first pressing temperature, 50 kgf compression force), as well as surface profiles of the microlens and the WC micromold cavity.

7.2.7   Measurement and Analysis of Optical Properties of Formed Glass Microlens Array

To verify the effectiveness of the proposed microthermal forming process for glass materials, the optical properties of the replicated glass microlens arrays were measured and analyzed. The average measured focal length for 10 replicated glass microlens samples was 249 μm (standard deviation: 6 μm) using a laser source with a wavelength of 609 nm. The calculated focal length was 246 μm using an optical simulation program (Code-V, Optical Research Associates). The measured and calculated focal lengths of the glass microlens were very similar. Figure 7.31 shows the measured light intensity profile of the replicated glass microlens and the diffraction-limited beam intensity profile at the focal plane. The focused beam spot was measured using a beam profiler and a laser source with a wavelength of

609 nm. A focused beam spot diameter of 4.3 μm at full-width half-maximum (FWHM) was obtained, and the diffraction-limited beam spot diameter (FWHM) was 3.5 μm.

REFERENCES

1. C. W. Sele, T. von Werne, R. H. Friend, and H. Sirringhaus (2005) Lithography-free, self-aligned inkjet printing with sub-hundred-nanometer resolution. *Advanced Materials* 17(8), 997–1001.

2. H. M. Nur, J. H. Song, J. R. G. Evans, and M. J. Edrisinghe (2002) Inkjet printing of gold conductive tracks. *Journal of Materials Science: Materials in Electronics* 13(4), 213–219.

3. A. L. Dearden, P. J. Smith, D. Y. Shin, N. Reis, B. Derby, and P. O'Brien (2005) A low curing temperature silver ink for use in inkjet printing and subsequent production of conductive tracks. *Macromolecular Rapid Communications* 26 (4), 315–318.

4. S. H. Ko, H. Pan, C. P. Grigoropoulos, C. K. Luscombe, J. M. Fréchet, and D. Poulikakos (2007) All-inkjet-printed flexible electronics fabrication on a polymer substrate by low-temperature high-resolution selective laser sintering of metal nanoparticles. *Nanotechnology* 18(34), 345202.

5. S. Y. Chou, P. R. Krauss, and P. J. Renstrom (1996) Imprint lithography with 25-nanometer resolution. *Science* 272(5258), 85–87.

6. W. J. Dauksher, N. V. Le, E. S. Ainley, K. J. Nordquist, K. A. Gehoski, S. R. Young, J. H. Baker, D. Convey, P. S. Mangat (2006) Nanoimprint lithography: templates, imprinting, and wafer pattern transfer. *Microelectronic Engineering* 83, 929–932.

7. S. Zankovych, T. Hoffmann, J. Seekamp, J.-U. Bruch, and C. M. Sotomayor Torres (2001) Nanoimprint lithography: challenges and prospects. *Nanotechnology* 12(2), 91–95.

8. G. Jung, E. Johnston-Halperin, W. Wu, Z. Yu, S. Wang, W. M. Tong, Z. Li, J. E. Green, B. A. Sheriff, A. Boukai, Y. Bunimovich, J. R. Heath, and R. S. Williams (2006) Circuit fabrication at 17 nm half-pitch by nanoimprint lithography. *Nano Letters* 6(3), 351–354.

9. J. Lim and S. Kang, (2009) Effect of process conditions on the properties of conductive tracks in direct imprinting and sintering of metal nano-powders, *Journal of Micromechanics and Microengineering*, 19, 125001 (7pp)

10. K. E. Petersen (1982) Silicon as a mechanical material. *Proceedings of the IEEE* 70(5), 420–457.

11. S. Kim, J. Kim, J. Lim, M. Choi, S. Kang, and S. Lee, H. Kim (2007) Nanoimprinting of conductive tracks using metal nanopowders. *Applied Physics Letters* 91(14), 143117-1–143117-3.

12. Y. Hirai, S. Yoshida, and N. Takagi (2003) Defect analysis in thermal nanoimprint lithography. *Journal of Vacuum Science and Technology B* 21(6), 2765–2770.

13. H. Becker and U. Heim (2000) Hot embossing as a method for the fabrication of polymer high aspect ratio structures. *Sensors and Actuators* 83, 130–135.

14. M. Bender, M. Otto, B. Hadam, B. Spangenberg, and H. Kurz (2002) Multiple imprinting in UV-based nanoimprint lithography- related material issues. *Microelectronic Engineering* 61–62, 407–413.

15. N. Lee, S. Choi, and S. Kang (2006) Self-assembled monolayer as an anti-adhesion layer on a nickel nanostamper in the nanoreplication process for optoelectronic applications. *Applied Physics Letters* 88, 073101.

16. S. Magdassi, A. Bassa, Y. Vinetsky, and A. Kamyshny (2003) Silver nanoparticles as pigments for water-based inkjet inks. *Chemistry of Materials* 15(11), 2208–2217.

17. J. Han, B. Min, and S. Kang (2008) Microforming of glass microlens array an imprinted and sintered tungsten carbide micromold. *International Journal of Modern Physics B* 22(31,32), 6051–6056.

18. S. M. Solonin and L. I. Chernyshev (1975) Effects of sintering conditions on the physicomechanical properties of porous tungsten. *Powder Metallurgy and Metal Ceramics* 14(10), 806–808.

19. Y. Aono, M. Negishi, and J. Takano (2000) Development of large aperture aspherical lens with glass molding. *Proceedings of SPIE* 4231, 16–23.

20. M. Katsuki and T. Kamano (2001) High-precision optical glass moulding, *Japan: Toshiba Machine Co. Ltd., High-Precision Machine Division.*

21. M. Zhou and B. K. A. Ngop (2003) Factors affecting form accuracy in diamond turning of optical components. *Journal of Materials Processing Technology* 138, 586–589.

22. W. B. Lee, D. Gao, C. F. Cheung, and J. G. Li (2003) An NC tool path translator for virtual machining of precision optical products. *Journal of Materials Processing Technology* 140, 211–216.

23. D. Daly (2001) *Microlens Arrays*, Taylor and Francis, New York.

24. S. Kang and S. Moon (2001) Design and fabrication of microoptical components for optical data storage by micromolding. *CLEO/Pacific Rim* 2, 16–17.

25. Z. D. Popovic, R. A. Sprague, and G. A. Neville Connell (1988) Technique for monolithic fabrication of microlens array. *Applied Optics* 27, 1281–1284.

26. S. Moon, N. Lee, and S. Kang (2003) Fabrication of a microlens array using a microcompression moulding with an electroformed mould insert. *Journal of Micromechanics and Microengineering* 13(1), 98–103.

27. M. Severi and P. Mottier (1999) Etching selectivity control during resist pattern transfer into silica for the fabrication of microlenses with reduced spherical aberration. *Optical Engineering* 38, 146–150.

28. K. Fujikawa, G. Hirakawa, T. Shiono, and K. Nomura (1997) Optical properties of a Si binary optic microlens for infrared ray, *10th International IEEE Micro Electro Mechanical Systems Conference*, 360–365.

29. H.-O. Andren (2001) Microstructures of cemented carbides. *Materials & Design* 22(6), 491–498.

30. R. Frykholm, B. Jansson, and H.-O. Andren (2002) The influence of carbon content on formation of carbo-nitride free surface layers in cemented carbides. *International Journal of Refractory Metals & Hard Material* 20, 345–353.

31. C. H. Allibert (2001) Sintering features of cemented carbides WC-Co processed from fine powders. *International Journal of Refractory Metals & Hard Material* 19, 53–61.

32. H. Yamaguchi and T. Shinmura (1999) Study of the surface modification resulting from an internal magnetic abrasive finishing process. *Wear* 225–229, 246–255.

33. S. Feygin, G. Kremen, and L. Igelstyn (1998) Magnetic-abrasive powder and method of producing the same. *US Patent 5846270*.

34. B. W. Ahn, W. B. Kim, S. J. Park, and S. J. Lee (2003) Ultra precision polishing of micro die and mould parts using magnetic field-assisted machining. *Proceedings of the Korean Society of Precision Engineering Conference*, 244–247.

35. G. Kleer and W. Doell (1997) Ceramic multilayer coatings for glass moulding applications. *Surface and Coatings Technology* 94–95, 647–651.

36. M. Hock, E. Schaffer, W. Doll, and G. Kleer (2003) Composite coating materials for the moulding of diffractive and refractive optical components of inorganic glasses. *Surface and Coatings Technology* 163–164, 689–694.

37. Y. Kojima (2005) Grinding technoloy of aspheric molds for glass-molding; Technical Digest. *Proceedings of the SPIE, TD03* pp. 44–46.

38. W. Choi, J. Lee, W. Kim, B. Min, S. Kang, and S. Lee (2004) Design and fabrication of tungsten carbide mould with micropatterns imprinted by microlithography. *Journal of Micromechanics and Microengineering*, 14(11), 1519.

CHAPTER 8

Micro/Nano-Optics for Light-Emitting Diodes

Light-emitting diodes (LEDs) have many advantages compared to conventional light sources: they are ecofriendly; have a long life, high color-rendering index, and low power consumption; and are durable and safe. Recently, the LED market has experienced strong growth, and it is forecast to reach over \$40 billion in 2015 [1–3]. By developing LED technologies, as shown in Figure 8.1, many types of LEDs with high luminance performance and high power, even those that are very small, have been manufactured. Production of high-brightness LEDs, and particularly high-power LEDs, supports new applications with considerable market potential. LEDs are becoming more popular in numerous applications such as electronic signs and signals, displays, mobile phone devices, automotive and traffic lights, and general illumination, as shown in Figure 8.2 [4,5].

Unfortunately, commercial simulation tools have been optimized for imaging optics rather than nonimaging optics such as illumination, so they cannot be used to design optical components and estimate optical properties. Therefore techniques for LED illumination design must be developed. With the current demand for LEDs, LED lenses must be fabricated in large quantities, and they must be durable and safe for a wide range of temperature and humidity. Generally, glass lenses are fabricated by grinding, polishing, direct machining, and glass molding. They have high optical performance and are safe for a wide range of temperature and humidity. But their fabrication methods are expensive, require a long cycle time, and have limited mass-production potential. Plastic lenses fabricated by injection molding are inexpensive, have a short cycle time, and can be mass-produced, but they have low optical performance compared to glass lenses and can be influenced by changes in temperature and humidity. Therefore, a new fabrication method for LED lenses is required.

Micro/Nano Replication: Processes and Applications, First Edition. Shinill Kang.
© 2012 John Wiley & Sons, Inc. Published 2012 by John Wiley & Sons, Inc.

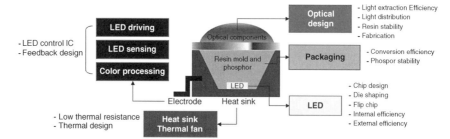

FIGURE 8.1 LED module technology issues.

In this chapter, LEDs are analyzed, and the problem of producing lenses for LED illumination is resolved by introducing micro-Fresnel lenses. Lenses with adequate optical performance are optimized. Then UV imprinting, which was discussed in Chapter 6, is applied as an alternative approach to fabricating micro-Fresnel lenses for LED illumination. These lenses are inexpensive, can be mass-produced, and have high optical and mechanical properties. The UV-imprinting process is optimized for the designed lenses. Finally, fabricated lenses are measured and analyzed to verify the feasibility of the UV-imprinting process.

8.1 DESIGNING AN INITIAL LENS SHAPE

8.1.1 LED Illumination Design

LEDs have various radiation profiles according to the structure of their packaging. To obtain a desirable optical performance for LED illumination,

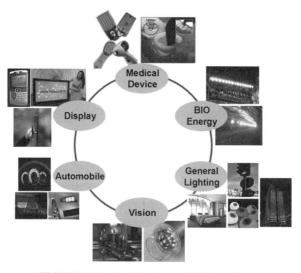

FIGURE 8.2 Applications of LED modules.

optical components are required because LEDs have limitations when they are the only source of illumination. The optical performance of illumination systems using only LEDs is debased because the viewing angle of LEDs is wide, which makes it hard to provide bright illuminate at the target area. In other words, LEDs have fundamental limits in many illumination applications because LED sources are ideal Lambertian emitter, which means that the intensity distribution I_θ is a cosine function of the emission angle. The flux efficiency of the original source with viewing angle θ_1 is therefore lower within the specific target area than that of the original source with an optical lens of viewing angle θ_2 because θ_1 is greater than θ_2. Here, flux efficiency is defined as the ratio of the incident flux within the target area, centered on the viewing angle θ_2, to the total flux emitted by the LED, and viewing angle is defined as the angle at which the luminous intensity is reduced to half its value at $0°$, as shown in Figure 8.3. The flux efficiency in the target area is easily improved by simple optical components with continuous surfaces, such as spherical lenses, because the intensity distribution of the original source becomes a power law of the cosine function. However, this causes the illuminance distribution E_θ to decrease more rapidly away from the optical axis [6].

Therefore, lenses were designed under the following considerations. First, it is difficult for LED illumination to illuminate intensively within a target area due to the wide radiation angle of the LEDs. Therefore, the lenses must collect the beam that emits beyond the target area to give the maximum luminance at the target area in order to improve the

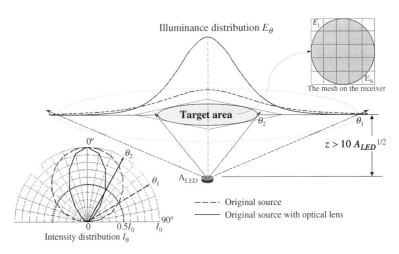

FIGURE 8.3   Schematic distribution of the intensity I_θ and the illuminance E_θ of an original source with and without an optical lens in the far-field distance. Reprinted with permission from Ref. [7]. Copyright 2009, Optics Express.

efficiency of the illumination system. Second, the lenses require uniform color distribution to remove the yellow ring phenomenon within the target beam angle and to provide a uniform luminance distribution at the illumination plane. Additionally, the lenses require small optical components for small LED illumination systems, such as a phone camera flash.

8.1.2 Source Modeling

An LED chip can be considered a point source aggregate. The beam radiated from each point source can be defined as a surface source. The beam radiates into a hemisphere according to the structure of the LED packaging. The inner structure of an LED was analyzed by cutting the LED. This allowed us to design the lenses, which control the illuminance distribution at the illumination plane. A white LED-based blue chip with a viewing angle of 60° was used in this study.

A white LED package consists of a blue LED chip, yellow phosphor, a reflector, and epoxy, as shown in Figure 8.4a. The white color is made by mixing the blue and yellow beams. Blue rays emitted from the chip are changed into white rays after passing through the yellow phosphor, as shown in Figure 8.4b. We assumed that the total beam from the LED was emitted from the top and side surfaces, totaling five surfaces. The rays emitted from the top surface pass through a thin layer with yellow phosphor, and the rays emitted from the side surface pass through a thick layer with yellow phosphor. As a result, there is a rich blue ray emission from the top surface and a rich yellow ray emission from the side surfaces [8].

The optical properties of LEDs were measured with a goniospectrometer and an integrating sphere. The intensity distribution, spectral distribution, and flux in all directions were measured. The LED was drawn with a CAD tool based on the measured geometrical data and designed by inputting the measured optical properties. Figure 8.5a illustrates the LED source modeling using a simulation tool. The analyzed simulation results of LED modeling, shown in Figure 8.5b, indicate that the relative intensity distribution for a white LED was directly proportional to the cosine of the angle from which it was viewed. The LED had a large beam divergence, and its radiation profile resembled a sphere. The emissions from the designed LED with a viewing angle of 60° also had a Lambertian distribution property.

8.1.3 Modeling a Spherical Refractive Lens

The first step in designing an optical lens for a given target viewing angle is finding the initial design values of the refractive surface. These determine the

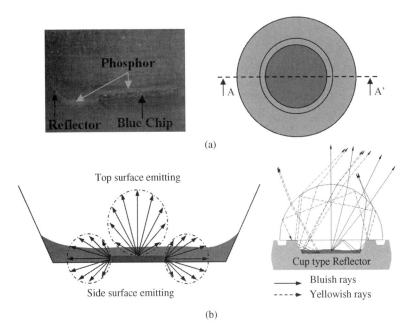

(a)

(b)

FIGURE 8.4 Analysis of a white LED module: (a) microscopic image at the surface of slice AA′; (b) main path of the rays emitted from each surface of the chip. Reprinted with permission from Ref. [7]. Copyright 2009, Optics Express.

optical power of the lens. For this step, we applied 2×2 ray matrices using derivatives between the direction and position of the ray vector emitted from the center of the LED source [9]. The initial radius of curvature of a normal spherical lens can be calculated using the geometric variables shown in Figure 8.6:

$$R = \frac{s n_L (n_L - 1) \tan \theta_{in} + t n_1 (n_L - n_1) \tan \theta_{in}}{n_L (\tan \theta_{out} - n_1 \tan \theta_{in})} \qquad (8.1)$$

where s is the distance between the first surface of the lens and the LED as determined from the geometry of the LED, θ_{in} is the viewing angle determined by the intensity distribution of the LED, θ_{out} is the target viewing angle for each illumination application, and n_1, n_L, and n_2 are the refractive indices of each medium. Once θ_{out} is determined, the radius of curvature with respect to t can be derived from Eq. (8.1), where t is the distance that an incident ray emitted from the LED source transmits between R_1 and R_2 in the direction of the optical axis Z, as shown in Figure 8.4. The total thickness T of the lens is approximately equal to t for paraxial rays.

(a)

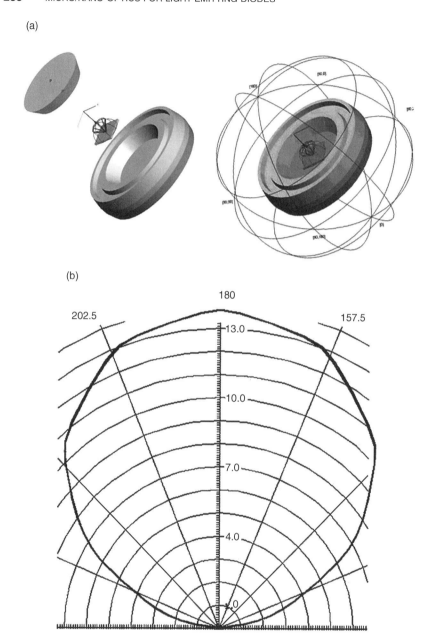

(b)

FIGURE 8.5 White LED (a) source modeling and (b) relative angular intensity.

8.1.4 Modeling a Micro-Fresnel Lens

Possible improvements in the optical power of a lens via traditional optical components, such as a spherical lens with a continuous refractive surface, are limited by constraints on the geometrical dimensions between the

(a)

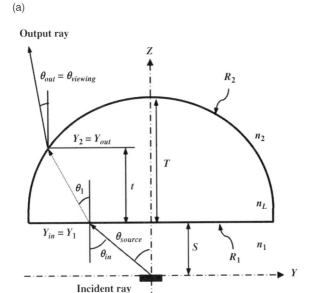

(b)

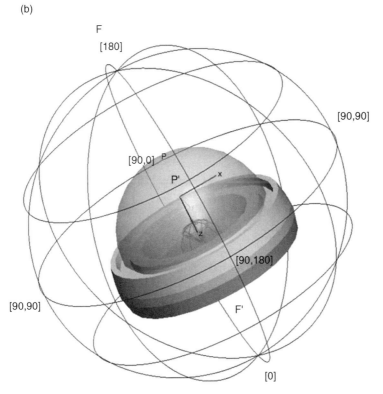

FIGURE 8.6   (a) Geometric configuration used to design a normal spherical lens. Reprinted with permission from Ref. [7]. Copyright 2009, Optics Express.

radius of curvature and the thickness for a fixed lens diameter. Therefore, micro-Fresnel lenses have become a viable option when designing optical components due to their geometric design freedom and low thickness. In a Fresnel lens design, the continuous refractive surface of traditional optics is divided and collapsed onto a bottom plane. Each discontinuous refractive surface connects to a groove facet that is generally parallel to the optical axis.

Figure 8.7 illustrates the design concept of a micro-Fresnel lens. When the groove angle is 0°, some rays transmitted through each discontinuous refractive surface deviate because of secondary refraction by the groove facet, which causes a large part of the optical loss in outer regions of the target area. Therefore, the angle of the groove facet can affect the optical efficiency and uniformity [10–12]. We drew an initial configuration of the micro-Fresnel lens using the pair values R and T, as shown in Figure 8.6. Our basic micro-Fresnel lens construction was such that the continuous refractive surface was divided into a number of equal heights, N. The profile of each discontinuous refractive surface $Z_n(y)$ and the positions y_n where the value of $Z_n(y)$ was zero were as follows:

$$y_n = \sqrt{R^2 - \left(R - \frac{nT}{N}\right)^2} \ (n = 1, 2, \ldots, N) \tag{8.2}$$

$$Z_n(y) = \frac{nT}{N} - \left(R - \sqrt{R^2 - y^2}\right), (y_{n-1} < y < y_n), \tag{8.3}$$

After the initial configuration of the discontinuous refractive surfaces was designed using Eqs. 8.2 and 8.3, each discontinuous refractive surface was connected to a modified groove facet and the base substrate of the micro-Fresnel lens, allowing the selection of h.

8.1.5   Verifying the Micro-Fresnel Lens Performance

Ray-tracing simulations for given practical conditions are necessary to calculate the optical power of the lenses. This is because the geometric variables of spherical and micro-Fresnel lenses are based on on-axis ray tracing, and the off-axis rays emitted from the LED are not considered [13,14].

We selected s, n_1, n_L, n_2, θ_{in}, and θ_{out} from Figure 8.6 as our fixed variables, choosing values of 1 mm, 1, 1.5, 1, 60°, and 30°, respectively.

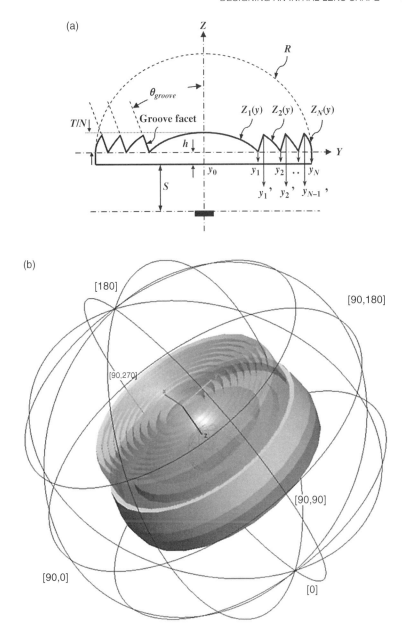

FIGURE 8.7 (a) Micro-Fresnel lens design concept used for the discontinuous surface and geometrical configuration used for the target viewing angle. Reprinted with permission from Ref. [7]. Copyright 2009, Optics Express.

Using these values in Eqs. (8.1)–(8.3), we performed a three-dimensional (3D) ray-tracing simulation of the entire optical system. We then corrected h and N in accordance with the target viewing angle. In this chapter, the

TABLE 8.1 Corrected Final Specifications of the Micro-Fresnel Lenses with Groove Angle of 0° for Each Target Viewing Angle

	Target Viewing Angle		
	20°	30°	40°
R (mm)	1.94	2.14	2.46
T/N (mm)	0.15	0.15	0.15
h (mm)	1.9	1.4	0.8

values of N were 17, 15, and 10 for target viewing angles 20°, 30°, and 40°, respectively. However, the groove height T/N may be altered due to fabrication issues. Table 8.1 shows the corrected specifications of the micro-Fresnel lenses according to the target viewing angle, and Figure 8.8 shows simulated intensity distributions for each case. The angles corresponding to half the normalized intensity at 0° matched the target viewing angles.

Figure 8.9 shows the simulated illumination distribution for different target areas of each viewing angle. The radius of the target area on the detecting plane was $z \cdot \tan \theta$, where z was 100 mm from the LED source to the detecting plane. As previously stated, the illumination curve follows a power cosine law. Thus, the smaller the target viewing angle was, the more rapidly the illumination distribution decreased on the same target plane.

The normalized deviation of illuminance F_1, which is a ratio of the standard deviation of illuminance to the average illuminance L_{avg}, and the normalized deviation of color F_2, which is a ratio of the standard deviation of color to the average color, were defined to evaluate the optical performance

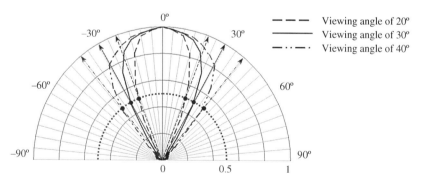

FIGURE 8.8 Simulated intensity distributions used to verify the corrected specifications of the micro-Fresnel lenses with a groove angle of 0° for each target viewing angle. Reprinted with permission from Ref. [7]. Copyright 2009, Optics Express.

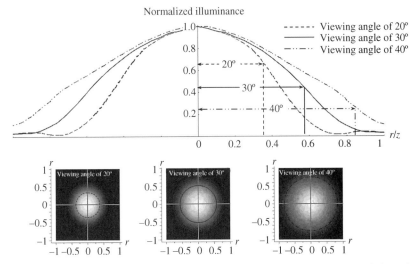

FIGURE 8.9 Simulated illuminance distributions of each micro-Fresnel lens in a normalized area, where r/z is 0.364, 0.577, and 0.839 for viewing angles of 20°, 30°, and 40°, respectively. Reprinted with permission from Ref. [7]. Copyright 2009, Optics Express.

of the micro-Fresnel lenses at the target plane for any target viewing angle [15–18]. Here, the functions F_1 and F_2 are as follows:

$$F_1 = \frac{1}{L_{avg}} \sqrt{\frac{\sum_{k=1}^{n} (L_k - L_{avg})^2}{n}} \tag{8.4}$$

$$F_2 = \frac{1}{\|(u_{avg}, v_{avg})\|}$$
$$\times \sqrt{\frac{\sum_{k=1}^{n} ((u_k - u_{avg})^2 + (v_k - v_{avg})^2)}{n}}, \left(\|(u_{avg}, v_{avg})\| = \sqrt{u_{avg}^2 + v_{avg}^2}\right) \tag{8.5}$$

where n is the total number of bins in the mesh of the detecting area, L_k is the illuminance of each bin, u_{avg} and v_{avg} are the average color coordinates for the total bins, and u_k and v_k are the color coordinates of each bin. The position of the detection plane was changed so that the areas of the bins were equal

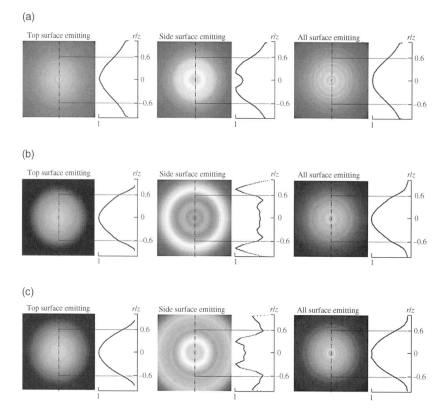

FIGURE 8.10 Comparison of color and normalized illuminance distributions of rays emitted from each surface of the LED source (a) without any lens, (b) with a spherical lens, and (c) with a micro-Fresnel lens. Reprinted with permission from Ref. [7]. Copyright 2009, Optics Express.

regardless of the target viewing angle. The detecting areas were divided into 25 bins considering the trade-off between resolution and accuracy in a statistical ray-tracing process.

The performance of the micro-Fresnel lenses was verified by comparing their behavior to that of spherical lenses. The color deviation observed from ray separation of a spherical lens with a viewing angle of 30° was markedly reduced by using a micro-Fresnel lens because the distribution of rays emitted from side surfaces of chips was scattered more moderately, as shown in Figure 8.10. The micro-Fresnel lens produced a more uniform color distribution with fewer deviations (0.00714 compared with 0.01761 for the spherical lens) for a target viewing angle of 30°, as shown in Figure 8.11.

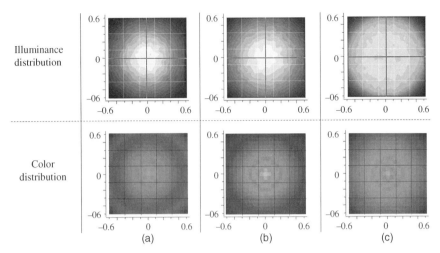

Illuminance distribution

Color distribution

(a)   (b)   (c)

FIGURE 8.11   Comparison of the normalized deviation of illuminance and color: (a) 0.334 and 0.01761 for a spherical lens and (b) 0.314 and 0.00714 for a Fresnel lens with groove angle of 0° and (c) 0.218 and 0.007 for a Fresnel lens with optimum groove angle of 29.4° at weight ratio of w1:w2=0.4:0.6. Reprinted with permission from Ref. [7]. Copyright 2009, Optics Express.

8.2 FABRICATION RESULTS AND DISCUSSION

8.2.1 Fabrication of the Micro-Fresnel Lens

When the size of a micro-Fresnel lens is large and it is required only in small quantities, it can be fabricated by direct machining. But replication methods are frequently required for mass production. UV-imprinting is one such method. It requires a master core with several groove tips in a concentric circle. The groove tips act as a refractive surface and affect the optical properties. Therefore, they must be fabricated with highly accurate geometries and surface roughness. The master core for each micro-Fresnel lens was fabricated by diamond-tool machining using a Nanoform-2000 machine, and then electroplated with nickel. The pattern of the fabricated master core was the inverse of the designed micro-Fresnel lens profile. Figure 8.12 shows an image of the fabricated master core.

UV imprinting consists of a chain of subprocesses, as shown in Figure 8.13. An UV-curable polymer was selected as the replication material [19]. During UV exposure, a relatively low pressure was applied to ensure good replication quality. The height of each groove increased with the compression pressure. The pressure applied during photopolymerization reduced the material shrinkage and the number of unfilled gaps. Therefore, complete filling of each groove cavity by an appropriate replication process was expected to prevent the illumination system from losing any optical efficiency.

FIGURE 8.12 Fabricated master core for a micro-Fresnel lens.

8.2.2 Elimination of Air Bubbles

Microair bubbles are frequently generated in the UV-imprinting process [20]. Two types of microair bubbles may occur: those that form inside the initial liquid photopolymer before any processing, and those that are generated at the interface between the photopolymer and the mold cavity by the interaction between the shape of the mold cavity and the viscosity of photopolymer. The latter case occurs during the photopolymer-coating process while the mold covers the substrate. These bubbles can cause various defects, resulting in optical losses in the optical components.

Microair bubbles can be avoided in two ways. One can use a vacuum system to evacuate the air from the photopolymer in a vacuum chamber. Microair bubbles in the initial liquid photopolymer can be easily eliminated in this way, but the method has a high initial cost and long cycle time. Alternatively, one can use a preheating process to raise the temperature of the photopolymer to around 40°C. The viscosity of the photopolymer decreases so that the microair bubbles become thermodynamically active. This allows

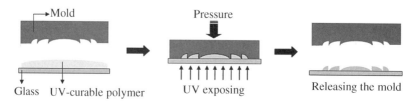

FIGURE 8.13 Process flow of UVimprinting for fabricating a modified micro-Fresnel lens. Reprinted with permission from Ref. [7]. Copyright 2009, Optics Express.

(a)

(b)

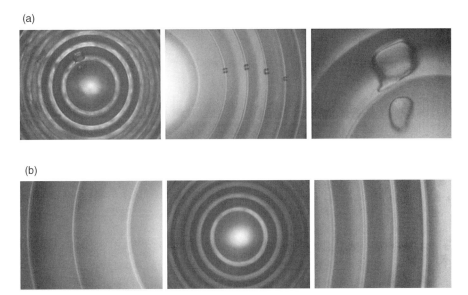

FIGURE 8.14 Image of a UV-imprinted micro-Fresnel lens (a) with an air bubble and (b) without an air bubble after preheating.

the photopolymer to be coated on a substrate and along the shape of the mold cavity without any microair bubbles forming during the photopolymer-coating process.

Due to the small and sharp tips and grooves of our micro-Fresnel lenses and the high viscosity of the photopolymer used in the UV-imprinting process, microair bubbles were frequently observed in the UV-imprinted parts. Figure 8.14a shows an image of a micro-Fresnel lens with an embedded microair bubble. Figure 8.14b shows an image of a micro-Fresnel lens without any microair bubbles obtained by preheating the polymer.

8.2.3 Optimization of the UV-Imprinting Process

It is important to transcribe the groove tips from the master in the replication process because they act as a refractive surface in a Fresnel lens. Poor replication of the groove tips can result in optical losses and deteriorate the optical performance of the lens. The UV-imprinting process can accurately transcribe optical components. However, the UV photopolymer shrinks during the photopolymerization, which deteriorates the replication quality. Thus, pressure was applied to compensate for the shrinkage [21,22]. The effects of varying the applied pressure were measured. Figure 8.15 shows that the sag height of the groove tip depended on the applied pressure at a fixed curing dose of 2000 mJ/cm^2. The height of the groove tip was close to that of

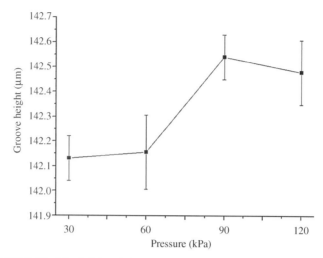

FIGURE 8.15 The height of the groove tip depends on the applied pressure.

the master groove tip, and the deviation between fabricated samples was minimized at an applied pressure of 90 kPa.

Details of the UV-imprinting process are as follows. The photopolymer was dispensed at the middle of the mold cavity by a dispenser. It slowly spread out over the blade between the concentric circles by compression while the mold covered the substrate. This helped to avoid air bubbles, even if the polymer was not preheated. Pressure was applied between the substrate and mold to compensate for photopolymer shrinkage. Figure 8.16 shows a surface image of each groove of the fabricated micro-Fresnel lens, and Figure 8.17 shows a fabricated lens using an optimum design and the accompanying LED module.

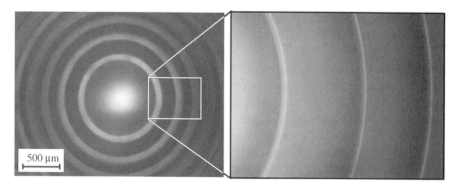

FIGURE 8.16 Microscopic image and surface profile of a fabricated modified micro-Fresnel lens with an optimum groove angle for a target viewing angle of 30°. Reprinted with permission from Ref. [7]. Copyright 2009, Optics Express.

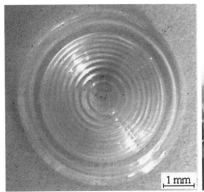 

FIGURE 8.17 Fabricated modified micro-Fresnel lens using an optimum design and the accompanying LED module. Reprinted with permission from Ref. [7]. Copyright 2009, Optics Express.

8.2.4 Evaluation of the Micro-Fresnel Lens for LED Illumination

The importance of having reliable measuring methods and equipment for LED development, testing, and certification has increased significantly in recent years. Based on the experiences of photometric laboratories and producers of LEDs, the International Commission on Illumination (CIE) provides helpful recommendations for measuring LED performance. CIE Technical Report 127/1 describes the measurement of photometric and radiometric quantities and introduces the concept of averaged LED intensity for two standard conditions [23]. Both conditions involve the use of a detector with a $100 \, \text{mm}^2$ circular-entrance aperture. The LED must be positioned facing the detector and aligned so that the mechanical axis of the LED passes through the center of the detector aperture. The distance between the LED tip and the entrance area of the detector varies between 316 mm (corresponding to a plane angle of $2°$) for condition A and 100 mm (corresponding to a plane angle of $6.5°$) for condition B. These definitions make it possible to compare different LEDs.

We measured the optical properties of the fabricated micro-Fresnel lenses, such as the relative angular intensity, luminance, and color coordinates, using a goniospectrometer, an integrating sphere, and a color analyzer based on CIE conditions A and B. Figure 8.18 shows the principles used to measure the optical properties.

A goniospectrometer, shown in Figure 8.19a, measures the angular intensity distribution and spectral distribution in three dimensions by turning the LED along its azimuthal and polar axes. An integrating sphere, shown in Figure 8.19b, is used for a variety of optical, photometric, or radiometric

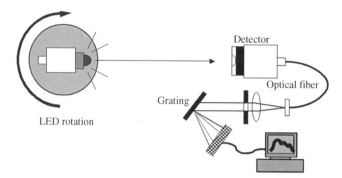

FIGURE 8.18 Schematic diagram of the principles used to measure the optical properties.

measurements. We used it to quantify the total light radiated in all directions from the LED. A color analyzer, shown in Figure 8.19c, measures the luminance and color coordinates. The main components of the color analyzer are an objective lens, an optical fiber block, on-chip lenses, and a sensor. The light from the light source is focused onto the receiving window of the optical fiber block. The focused light is mixed inside the block and split into three parts, which are then guided to the receiving areas of the x, y, and z sensors. The light is further focused by the on-chip lenses onto the sensors themselves.

The results of simulations and measurements were compared and analyzed to evaluate the feasibility of the UV-imprinting process for fabricating lenses for LED illumination. A diffuser plate, 120×120 mm, was placed 100 mm above the LED source to measure the illuminance and color distribution quality. The diffuser plate was divided into 25 detecting regions to calculate

(a)

FIGURE 8.19 The system of (a) Goniospectrometer, (b) integrating sphere, and (c) color analyzer.

(b)

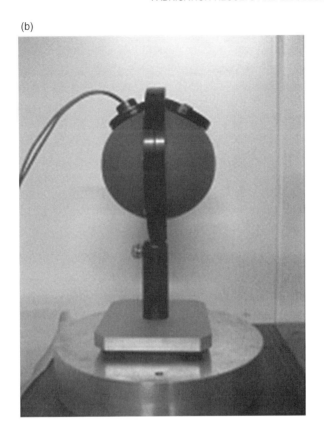

(c)

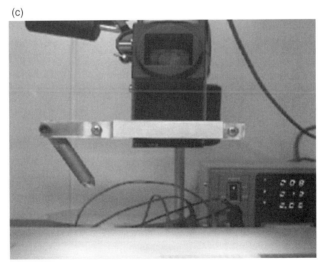

FIGURE 8.19 (*Continued*)

the normalized deviation of illuminance and color. For a target viewing angle of 30°, the measured values of the normalized deviations of illuminance and color were 0.331 and 0.0214 for a spherical lens, and 0.214 and 0.0079 for the fabricated micro-Fresnel lens, respectively, as shown in Figure 8.20b. The measured and simulated results were in good agreement.

(a)

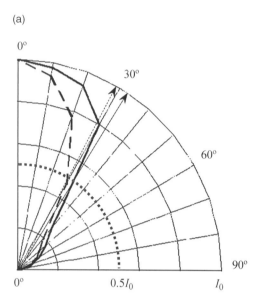

(b)

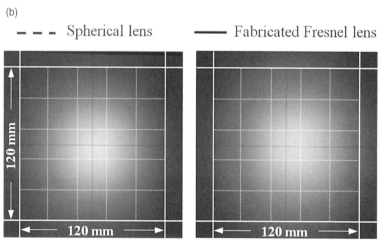

FIGURE 8.20 Measured (a) intensity distributions of each lens for a target viewing angle of 30° and (b) distribution of illuminance and color with normalized deviation of illuminance and color of 0.331 and 0.0214 for the spherical lens and 0.214 and 0.0079 for the fabricated micro-Fresnel lens. Reprinted with permission from Ref. [7]. Copyright 2009, Optics Express.

8.3 CONCLUSIONS

Micro-Fresnel lenses were designed to meet the increasing needs and requirements of lenses for LED illumination. General spherical lenses lose their flux efficiency and illuminance uniformity due to separation between blue and yellow rays. Therefore, our micro-Fresnel lenses were designed to minimize the optical loss and maximize the color uniformity. The initial optical power of the micro-Fresnel lenses was derived from 2×2 ray matrices and verified by a ray-tracing simulation. Other optical effects of the micro-Fresnel lenses were verified analytically and experimentally. Lenses for viewing angles of $20°$, $30°$, and $40°$ were derived based on each optical function. An UV-imprinting process was selected and developed to fabricate the lenses, and the feasibility of the process was analyzed by measuring the resulting geometrical and optical properties. The following conclusions may be drawn.

(1) A white LED-based blue chip with a viewing angle of $60°$ was used as the target LED of this study. The light source from an LED was analyzed to design appropriate lenses for LED illumination. The optical and geometrical properties were measured to simulate the LED, and spherical refractive lenses were designed as references. Then, micro-Fresnel lenses were designed and optimized to provide high luminance to improve the efficiency of the illumination system within the illumination plane. The color deviation was reduced due to moderate mixing between the blue and yellow rays emitted from the LED by each refractive surface of the micro-Fresnel lenses.

(2) Possible replication methods for the designed micro-Fresnel lenses were resin transfer molding and UV imprinting. We selected the UV-imprinting process. The master core for the UV-imprinting process was fabricated by diamond-tool machining. Through analysis, the resulting photopolymerization, air bubbles, and photopolymer shrinkage were reduced to improve the optical properties of mass-produced LEDs. The curing dose for photopolymerization was calculated by using the Fourier transform infrared (FTIR) spectroscopy. The air bubbles were eliminated by preheating the photopolymer. Photopolymer shrinkage was compensated by applying a pressure of 90 kPa.

(3) The feasibility of an UV-imprinting process for fabricating micro-Fresnel lenses was evaluated. The optical properties of the fabricated micro-Fresnel lenses were measured and compared with simulation results. The fabricated micro-Fresnel lenses met the requirements for

LED illumination. The luminance within the target viewing angle was increased, and the color distribution was uniform, removing the yellow ring phenomenon that can be generated by spherical refractive lenses. In addition, the thickness of the micro-Fresnel lenses was less than that of the corresponding spherical lenses.

REFERENCES

1. Y. Uchida and T. Taguchi (2005) Lighting theory and luminous characteristics of white light-emitting diodes. *Optical Engineering*, 44, 124003–1.

2. S. Bierhuizen, M. Krames, G. Harbers, and G. Weijers (2007) Performance and trends of high-power light-emitting diodes. *Proceedings of SPIE*, 6669, 66690B-1–66690B-12.

3. Global Information, Inc. *Lighting and LED Market Research*, http://www.the-infoshop.com/topics/EL05_en.shtml.

4. D. A. Steigerwald, J. C. Bhat, D. Collins, R. M. Fletcher, M. O. Holcomb, M. J. Ludowise, P. S. Martin, and S. L. Rudaz (2002) Illumination with solid state lighting technology. *IEEE*, 8, 310–320.

5. P. Manninen, J. Hovila, P. Karha, and E. Ikonen (2007) Method for analysing luminous intensity of light-emitting diodes. *Measurement Science and Technology*, 18, 223–229.

6. I. Moreno, M. Avendaño-Alejo, and R. I. Tzonchev (2006) Designing light-emitting diode arrays for uniform near-field irradiance. *Applied Optics*, 45, 2265–2272.

7. B.Kim, M. Choi, H. Kim, J. Lim, and S. Kang (2009) Elimination of flux loss by optimizing the groove angle in modified Fresnel lens to increase illuminance uniformity, color uniformity, and flux efficiency in LED illumination. *Optics Express*, 17(20), 17916–17927.

8. Á. Borbély and S. G. Johnson (2005) Performance of phosphor-coated light-emitting diode optics in ray-trace simulations. *Optical Engineering*, 44, 111308.

9. F. L. Pedrotti, S. J, Leno, M. Pedrotti, and L. S. Pedrotti, (2007) *Introduction to Optics*, 3rd Edition, Pearson, ST, San Francisco.

10. O. E. Miller, J. H. Mcleod, and W. T. Sherwood (1951) Thin-sheet plastic micro-Fresnel lenses of high aperture. *Journal of the Optical Society of America A*, 41, 807–815.

11. W. Watanabe, D. Kuroda and K. Itoh (2002) Fabrication of Fresnel zone plate embedded in silica glass by femtosecond laser pulse. *Optics Express*, 10, 978–983.

12. C. Sierra and A. J. Vazquez (2005) High solar energy concentration with Fresnel lens. *Journal of Materials Science*, 40, 1339–1343.

13. S. Lee (2001) Analysis of light-emitting diodes by Monte Carlo photon simulation. *Applied Optics*, 40, 1427–1437.

14. C. Sun, T. Lee, S. Ma, Y. Lee, and S. Huang (2006) Precise optical modeling for LED lighting verified by cross-correlation in the midfield region. *Optics Letter*, 31, 2193–2195.

15. G. Wyszecki and W. S. Stiles (1982) *Color Science*, 2nd Edition, Wiley, New York.

16. F. Fournier and J. Rolland (2008) Optimization of freeform light pipes for light-emitting-diode projectors. *Applied Optics*, 47, 957–966.

17. Y. Zhen, Z. Jiaa, and W. Zhang (2007) The optimal design of TIR lens for improving LED illumination uniformity and efficiency. *Proceedings of SPIE*, 6834, 68342K-1–68342K1-8.

18. I. Moreno and U. Contreras (2007) Color distribution from multicolor LED arrays. *Optics Express*, 15, 3607–3618.

19. J. Lim, K. Jeong, J. Yoo, N. Park, and S. Kang (2005) Design and fabrication of diffractive optical element for objective lens of small form factor optical data storage device. *Information Storage and Processing System Conference*.

20. P. Nussbaum, I. Philipoussis, A. Husser, and H.P. Herzig (1998) Simple technique for replication of micro-optical elements. *Optical Engineering*, 37, 1804–1808.

21. S. Kim and S. Kang (2003) Replication qualities and optical properties of UV-moulded microlens arrays. *Journal of Physics D: Applied Physics*, 36, 2451–2456.

22. H. Kim, J. Lee, J. Lim, S. Kim, and S. Kang (2006) Design of microlens illuminated aperture array fabricated by aligned ultraviolet imprinting process for optical read only memory card system. *Applied Physics Letters*, 88, 241114.

23. Dr.-I. T. Q. Khanh (2003) *Radiometric and Photometric Measuring Systems for LEDs*, Arnold & Richter AG, Department R&D.

CHAPTER 9

Micro-/Nano-Optics for Optical Communications

In our digital and information-oriented society, the market size of optical components has being growing rapidly because the quantity of information used by people has sharply increased in areas such as optical communication, optical sensors, optical data storage devices, and digital displays. Recently, an interest has emerged in optoelectronic modules with integrated electronic micro-optical components [1]. In these modules, a microlens array is placed on the core component to improve the optical efficiency and resolution. Laser-to-fiber or fiber-to-fiber coupling modules are used to adjust and couple the beam generated from a semiconductor laser or emitted from an optical fiber, which is one of the key elements in optical fiber networks. Currently, laser-to-fiber and fiber-to-fiber couplings are predominantly used to transmit data signals in the optical communication industry. Although existing commercial coupling modules have high coupling efficiency, they commonly require time-consuming fabrication processes and expensive external robotic controls to adjust the laser-to-fiber or fiber-to-fiber alignment. One of the recent trends in laser-to-fiber and fiber-to-fiber alignment has been the use of silicon-platform technology, where silicon wafers with V-grooves for the fiber attachment and a marked area for precision alignment act as an optical bench [2–5]. This alignment technique needs precision control. A fiber ferrule with several multi-mode or single-mode optical fibers is often used. To further improve the coupling efficiency to 30–80%, a microlens array can be placed between the source array and the optical fiber. In this study, the laser source was a vertical-cavity surface-emitting laser (VCSEL) array. The VCSEL is a semiconductor laser with a laser beam emission perpendicular to the chip surface, contrary to conventional edge-emitting semiconductor lasers (also in-plane lasers) in which the laser light is emitted at one or two edges.

Micro/Nano Replication: Processes and Applications, First Edition. Shinill Kang.
© 2012 John Wiley & Sons, Inc. Published 2012 by John Wiley & Sons, Inc.

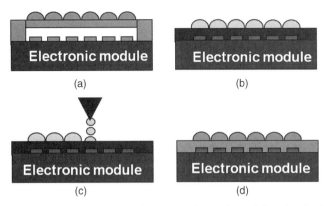

FIGURE 9.1   Integrating the microlens for an optoelectronic module using the (a) align-and-assemble method, (b) reflow method, (c) microdrop method, and (d) integrated method.

The VCSEL has many advantages, including low fabrication cost, low power consumption, on-wafer testability, high coupling efficiency with optical fibers, and array producibility. It can be used with single-mode lasers, is wavelength tunable, and has high intrinsic modulation bandwidths and low emission power compared with edge-emitting lasers [6]. Microlenses have been manufactured using this laser-to-fiber alignment and the align-and-assemble method [7,8], the reflow method [9], the microdrop method [10], and the integrated method [11], as shown in Figure 9.1.

In this chapter, two types of production methods for microlens arrays are examined: microcompression molding [12,13] and micro-UV-imprinting [11]. Microcompression molding was used to produce separated microlens arrays, which must then be aligned and assembled before being used to improve the coupling efficiency of the laser source in a VCSEL array. Separated microlens arrays were produced because they can be used to couple a VCSEL array with a multi-mode step-index optical fiber. A multi-mode step-index optical fiber has a core diameter that is 10 times greater than that of a coupled single-mode step-index optical fiber, which is expensive and time-consuming to couple with a laser source. Integrated microlens arrays for coupling a VCSEL array with a single-mode step-index optical fiber were also designed and produced using micro-UV-imprinting. A UV-transparent mold was fabricated for this process, and the microlens arrays were integrated directly on the VCSEL arrays. During the design stages of the two types of microlens arrays, the design variables were optimized to maximize the optical fiber coupling. A Monte Carlo analysis was performed to determine the critical tolerances based on yield and reliability requirements. The resulting array surface profiles were measured and compared with the designed surface profiles to confirm that the microlens arrays were produced to the desired specifications. Finally, a system was designed to measure the coupling efficiency between the VCSEL arrays and the optical fibers through the microlens arrays. The coupling efficiency measurements indicated that the performance of the microlens arrays was acceptable.

9.1 FIBER COUPLING THEORY

The basic coupling scheme is shown in Figure 10.3. The light beam emitted from a practical laser diode is focused on the optical fiber. The mode fields of the optical fiber are solutions to the following equation [14]:

$$\frac{d^2\phi_f}{dr^2} + \frac{1}{r}\frac{d\phi_f}{dr} + \left\{ \frac{n_1 k^2 \left(1 - 2\sqrt{NA/n_1}\right)(2p+l+1)}{a} - \frac{l^2}{r^2} \right\}\phi_f = 0 \quad (9.1)$$

where ϕ_f is the mode field of the fiber, r is the radial direction coordinate, n_1 is the refractive index of the fiber core, a is the core diameter of the optical fiber, k is the propagation constant for light in a vacuum, NA is the numerical aperture of the optical fiber, and p and l are the radial and rotational mode numbers, respectively. The total number of guided modes is given by the following:

$$M = \frac{V^2}{2} \quad (9.2)$$

where V is the normalized frequency for the fiber, defined as follows:

$$V = \frac{2\pi}{\lambda}aNA = \frac{2\pi}{\lambda}a\sqrt{n_1^2 - n_2^2} \quad (9.3)$$

where λ is the free-space wavelength. However, at surface 3, the mode field of the light emitted from the laser is [15] as follows:

$$\phi_{L,3} = \left(\sqrt{2}\frac{r}{w_{L,3}}\right)^l \cdot L_p^l\left(2\frac{r^2}{w_{L,3}^2}\right) \quad (9.4)$$

where L_p^l is a Laguerre polynomial, $w_{L,3}$ is the waist spot diameter at surface 3, which can be calculated using $ABCD$ matrix analysis [16] from surface 0 (the laser diode) to surface 3, and p and l are the radial and angular mode numbers, respectively. The Laguerre polynomial $L_p^l(x)$ obeys the following differential equation:

$$x\frac{d^2 L_p^l}{dx^2} + (l+1-x)\frac{dL_p^l}{dx} + pL_p^l = 0 \quad (9.5)$$

The coupling efficiency between the laser diode and the optical fiber through the lens is given by the well-known overlap integral [17,18]:

$$\eta = \left| \int_{S_3} \phi_{L,3} \cdot \phi_f dS_3 \right|^2 = 4\pi^2 \left| \int_0^{+\infty} \phi_{L,3} \cdot \phi_f r dr \right|^2 \qquad (9.6)$$

9.2 SEPARATED MICROLENS ARRAY

9.2.1 Design

The separated microlens array was designed to couple the multi-mode step-index optical fiber and the VCSEL array. Multi-mode step-index fibers have relatively large core diameters and large numerical apertures. A large core size and a large numerical aperture make it easier to couple light from a VCSEL into the optical fiber. The basic concept of the separated microlens array and the design variables are shown in Figure 9.2. Light emitted by a laser source is transmitted through the separated microlens array and focused on the optical fiber array. Here, D, R, and x_2 are the diameter, radius of curvature, and thickness of the microlens, respectively, x_1 is the distance from the light source to the front surface of the microlens, known as the front conjugate distance, and x_3 is the distance from back surface of the microlens to the optical fiber array, known as the back conjugate distance.

The light source was a 12-channel VCSEL array with a 9.5° semi-divergence angle and a laser beam passing through a 12-channel multi-mode step-index optical fiber ferrule with natural numerical aperture of 0.165.

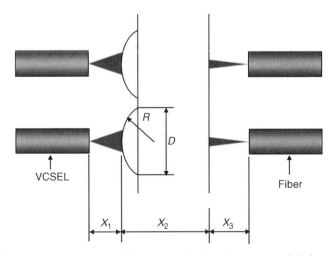

FIGURE 9.2   Design concept of the separated microlens array and design variables.

TABLE 9.1 Light Source Specifications for the Separated Microlens Array

Light Source	12-Channel VCSEL Array
Semi-divergence angle	9.5°
Numerical aperture	0.165
Diameter of emitting area	18 μm
Pitch of each source	250 μm

TABLE 9.2 Multi-mode Optical Fiber Specifications for the Separated Microlens Array

Numerical aperture of fiber	0.165
Diameter of fiber core	50 μm
Diameter of cladding	125 μm
Pitch of fiber array	250 μm

Each light source had a wavelength of 850 nm and a pitch of 250 μm. The specifications of each light source and receiving fiber ferrule are listed in Tables 9.1 and 9.2. The design variables were optimized considering the manufacturing and assembly tolerances for the optimum design state. The insertion loss was −0.86 dB. The design variables and optimized values are given in Table 9.3. The desired insertion loss was less than −3.0 dB, with a tolerance of ±2.0° in tilt, ±20 μm in radius of curvature, ±10 μm in front conjugate distance, ±20 μm in thickness of the microlens, ±5 μm in decenter, and ±20 μm in back conjugate distance.

9.2.2 Fabrication

The separated microlens array was replicated using microcompression molding. The mold insert for the microlens array was fabricated using the reflow [9] and electroforming [8] methods. First, photoresist islands 230 μm in diameter and 250 μm in pitch were made on a silicon substrate with a thickness of 500 μm and a diameter of 100 mm using a standard

TABLE 9.3 Variables and Optimum Values for the Separated Microlens Array

Front conjugate distance	350 μm
Thickness of microlens	780 μm
Back conjugate distance	180 μm
Diameter of microlens	230 μm
Radius of curvature of microlens	150 μm

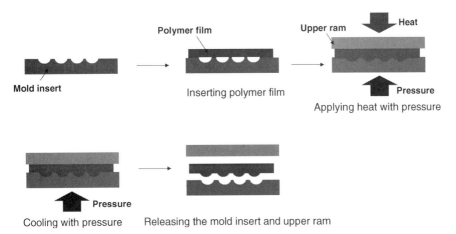

FIGURE 9.3 Process flow of microcompression molding.

photolithographic processing sequence. Hoechst AZ4620 was used as the photoresist material. Then, the photoresist islands were heated on a hot plate. The surface tension formed the photoresist islands into lens shapes. A metallic mold insert was fabricated using electroforming on the reflow lens array. An electron beam evaporation system was used to deposit a nickel seed layer on the mother, after which the nickel was electroformed. Commercial nickel sulfamate solution was used as an electrolyte. The electroformed nickel was used as a metallic mold insert to replicate the microlens array using microcompression molding.

The microcompression molding system was modified from a conventional macrosystem to improve the transcribability of the micro- to nanometer features. Figure 9.3 illustrates the microcompression molding process, which was similar to general macrocompression molding and consisted of inserting the polymer film and heating, pressing, and cooling the mold while at first holding and then releasing the pressure. The mold was heated above the softening temperature of the polymer film, and a small prepressure was applied to maintain contact between the melting polymer film and the mold insert. Compression pressure was applied so that microlens array could be replicated when the mold temperature exceeded the softening temperature. The mold was then cooled while the compression pressure was maintained. It was important to use the proper cooling rate to assure the quality of the microlens array. Once the demolding temperature had been reached, the replicated microlens array was released from the metallic mold insert. Figure 9.4 shows a schematic diagram of microcompression molding system, and Figure 9.5 shows scanning electron microscopy (SEM) images of the fabricated mold insert and replicated microlens array.

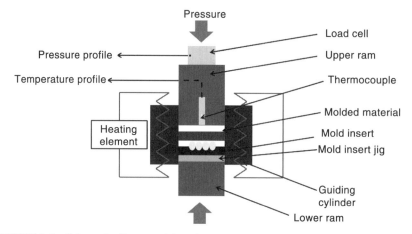

FIGURE 9.4 Schematic diagram of the microcompression molding system. Reprinted with permission from Ref. [19]. Copyright 2008, World Scientific.

(a)

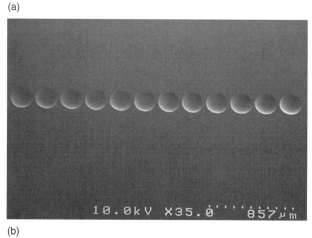

(b)

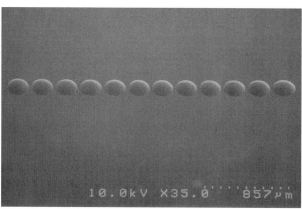

FIGURE 9.5 SEM images of the (a) fabricated mold insert and (b) replicated microlens array.

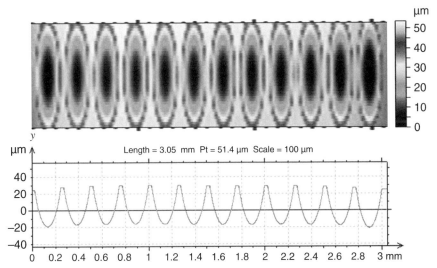

FIGURE 9.6   Surface shape image of the fabricated mold insert obtained using Form Talysurf.

9.2.3 Measurement Results

The produced microlens array was measured to confirm that it had the desired surface profile. Figure 9.6 shows the measured surface shape of the fabricated mold insert obtained using Form Talysurf. Figure 9.7 compares the section profiles of the designed separated microlens and the fabricated mold insert. Figures 9.6 and 9.7 illustrate that the fabricated mold insert had an average

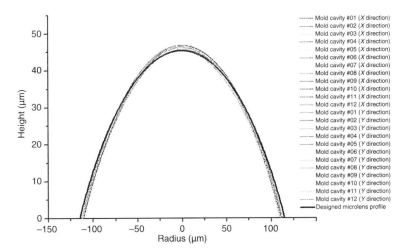

FIGURE 9.7   Comparison between surface profiles of the designed separated microlens array and the fabricated mold insert.

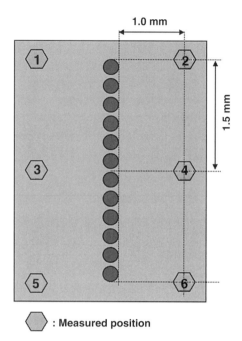

FIGURE 9.8 Substrate thickness measurement position for the replicated microlens array.

diameter of 226.25 μm (standard deviation: 1.21 μm), an average sag height of 46.45 μm (standard deviation: 0.49 μm), a pitch of 250 μm, and a surface roughness (R_t) of 2.31 nm. Figure 9.8 and Table 9.4 give the position where the substrate thickness was measured using a height gauge (accuracy: 1 μm at 20°C) and an XY positioner, along with the measured values. The measured mean value of the substrate thickness was 735.7 μm, which was larger than optimum design value (734.6 μm), but was still deemed acceptable, as the substrate thickness tolerance was ±20 μm.

Figure 9.9 shows the measured surface shape of the replicated microlens array using Form Talysurf. Figure 9.10 compares the section profiles of the

TABLE 9.4 Substrate Thickness of the Separated Microlens Array

Measurement Position	Measured Substrate Thickness
1	735.2 μm
2	734.7 μm
3	735.8 μm
4	735.5 μm
5	736.3 μm
6	736.6 μm

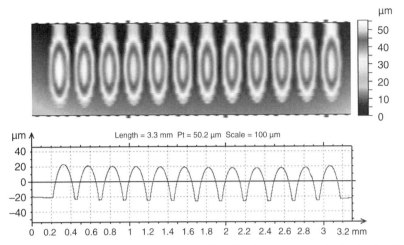

FIGURE 9.9 Surface shape image of the replicated microlens array obtained using Form Talysurf.

separated microlens and the replicated microlens array. Figures 9.9 and 9.10 indicate that the replicated microlens array had an average diameter of 225.89 μm (standard deviation: 0.77 μm), a sag height of 45.53 μm (standard deviation: 0.23 μm), a pitch of 250 μm, and an R_t of 1.22 nm. Finally, the insertion loss of the microlens array was measured. The intensity of the VCSEL array at a wavelength of 850 nm was measured through the optical fiber and set as a reference. The intensity of the light from the VCSEL through the microlens and the optical fiber was also measured. The coupling efficiency was calculated using these two values. The insertion loss of the microlens, which was a

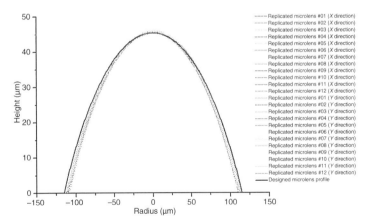

FIGURE 9.10 Comparison between surface profiles of the designed separated microlens array and the replicated microlens array.

TABLE 9.5 **Measured Insertion Loss of the Separated Microlens Array**

Channel	Insertion Loss
1	−0.86 dB
2	−0.50 dB
3	−0.73 dB
4	−0.55 dB
5	−0.91 dB
6	−1.12 dB
7	−1.72 dB
8	−2.03 dB
9	−2.68 dB
10	−1.94 dB
11	−2.39 dB
12	−1.32 dB

connector between the VCSEL and fiber, was expressed as follows:

$$F = 10 \log (I_o/I_i) \tag{9.7}$$

where I_i and I_o are the light intensity of the emitting area of the VCSEL and the detecting area of the fiber, respectively. The measured insertion loss is shown in Table 9.5. The average value and standard deviation of the insertion loss were −1.49 and 0.74 dB, respectively. The mean insertion loss was less than −3 dB, which was the design specification.

9.3 INTEGRATED MICROLENS ARRAY

9.3.1 Design

The integrated microlens array was designed to couple the single-mode step-index optical fibers and the VCSEL array. Single-mode step-index fibers propagate only one mode, called the fundamental mode. These fibers have low-attenuation and high-bandwidth properties. Present applications for single-mode fibers include long-haul high-speed telecommunications. The integrated microlens array was produced to couple the optical fiber and VCSEL array. The microlens array was aligned and fabricated directly on the VCSEL array. The basic concept of the integrated microlens array and the design variables are shown in Figure 9.11. Light emitted by a source is transmitted through the integrated microlens array and focused on the optical fiber array. Here, D and R are the diameter and radius of curvature of the microlens, respectively, z_1 is the thickness of the polymer substrate, z_2 is the microlens sag height, and z_3 is the distance from the surface of the microlens to the optical fiber array.

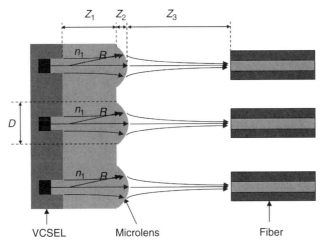

FIGURE 9.11 Design concept of the integrated microlens array and design variables.

The light source was a 30×30 VCSEL array with a $14°$ semi-divergence angle and a laser beam passing through a 30×30-channel single-mode optical fiber with a natural numerical aperture of 0.13. Each light source had a wavelength of 850 nm. The specifications of the light source and receiving optical fiber are listed in Tables 9.6 and 9.7. The design variables were optimized by considering the manufacturing and assembly tolerances and for the optimum design. The insertion loss was -0.106 dB. The design variables and optimized values are listed in Table 9.8. Figure 9.12 shows the emitted light from the VCSEL through the designed microlens, which had a diameter of 40 μm, a sag height of 2.52 μm, a focal length of 150 μm, and a polymer substrate thickness of 20 μm. The desired insertion loss was less than -3.0 dB, with a tolerance of $\pm 1.5°$ in tilt, ± 2 μm in decenter, and ± 10 μm in lateral direction spacing.

TABLE 9.6 Light Source Specifications for the Integrated Microlens Array

Light source	30×30 VCSEL array
Semi-divergence angle	$7°$
Numerical aperture	0.122
Diameter of emitting area	18 μm

TABLE 9.7 Single-Mode Optical Fiberspecifications for the Integrated Microlens Array

Numerical aperture of fiber	0.13
Diameter of fiber core	5 μm
Diameter of cladding	125 μm

TABLE 9.8 Design Variables and Optimum Values for the Integrated Microlens Array

Thickness of polymer substrate	20 μm
Sag height of microlens	2.52 μm
Distance from surface of microlens to fiber	150 μm
Diameter of microlens	40 μm
Radius of curvature of microlens	80.63 μm

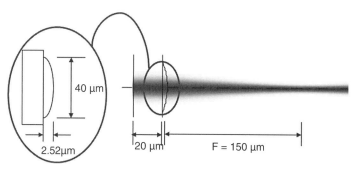

FIGURE 9.12 Emitted light from the VCSEL through the designed integrated microlens array.

9.3.2 Fabrication

The micro-UV-imprinting process molds an optical component on an electronic component while partially patterning the surface at the same time. The processing illustrated in Figure 9.13 was used to produce the integrated microlens array. First, the photopolymer was poured, covering the transparent mold, and aligned with the electronic module. Then the photopolymer was polymerized by UV exposure. A finished product was made after demolding

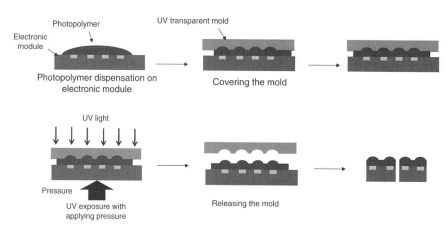

FIGURE 9.13 Process flow of micro-UV imprinting for an integrated microlens array.

FIGURE 9.14 SEM image of the UV-transparent mold.

the UV-transparent mold and dicing the product into the desired numbers of modules. Therefore, micro-UV-imprinting systems require a basic UV light source and collimating optical system. An alignment system is also required to align the mold and the electronic component, and a pressing system is needed to compensate for the polymer shrinkage [11]. The UV-transparent mold made for the designed microlens array mold is shown in the SEM image in Figure 9.14. Micro-UV-imprinting was then performed on a VCSEL array chip using a UV photopolymerized polymer belonging to the acrylate series with a viscosity of 400 cps. The UV-imprinting process took place at a radiation intensity of 2000 mJ under a pressure of 90 kPa. Figure 9.15 shows the bare VCSEL array and the integrated microlens array produced on the VCSEL array.

9.3.3 Measurement Results

It is necessary to verify that the desired surface was produced on the microlens array. Figures 9.16a and b show the measured three-dimensional

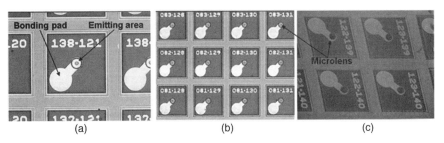

FIGURE 9.15 (a) Microscope image of a bare VCSEL; (b) microscope image; (c) SEM image of an integrated microlens array on the VCSEL array.

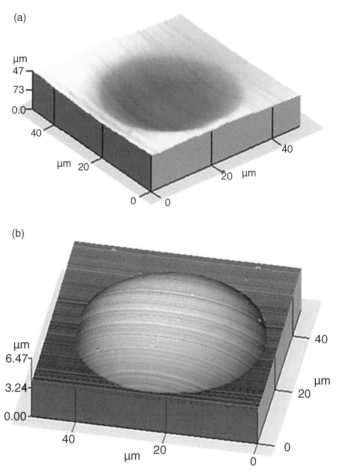

FIGURE 9.16 AFM three-dimensional profile of the (a) fabricated UV-transparent mold and (b) replicated microlens.

profiles of the fabricated UV-transparent mold and the replicated integrated microlens obtained using atomic force microscopy (AFM). Figure 9.17 compares the section profile of the designed integrated microlens and the microlens cavity of the fabricated UV-transparent mold. The fabricated UV-transparent mold had a diameter of 50.28 µm, a sag height of 2.53 µm, and a surface roughness (R_{rms}) of 2.47 nm. This was larger than the designed diameter of 40 µm. Figure 9.18 compares the section surface profiles of the designed microlens and the produced integrated microlens arrays. The produced microlens array had a diameter of 42.13 µm and a sag height of 2.51 µm. The error between the designed value and the actual product was less than 5 µm. The actual R_{rms} was 1.24 nm, and the actual total height, consisting of the sum of the polymer substrate thickness and the microlens

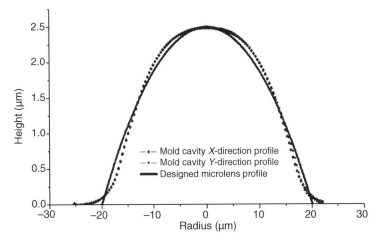

FIGURE 9.17 Comparison between surface profiles of the designed integrated microlens and the reversed surface profile of the fabricated UV-transparent mold.

sag height, was 26.34 μm at a pitch of 330 μm. The designed height was 22.52 μm with a tolerance of ±5 μm. Therefore, the dimensions of the replicated microlens were acceptable.

A measurement system that could measure the coupling efficiency of the VCSEL array with an integrated microlens array was designed to measure the insertion loss, as shown in Figure 9.19. The coupling efficiency was measured using a test board with a power supply, a three-axis stage to align the test

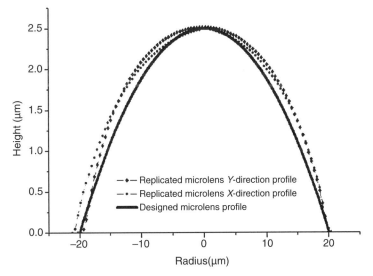

FIGURE 9.18 Comparison between surface profiles of the designed integrated microlens array and the replicated microlens array.

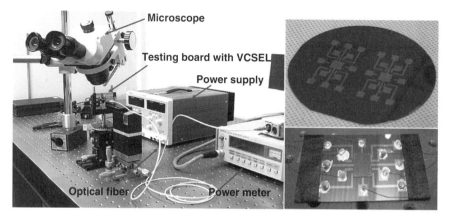

FIGURE 9.19 (a) Insertion loss measurement system for the integrated microlens array; (b) testing board with and without the VCSEL array.

board with the optical fiber, and a power meter to measure the intensity of radiation inserted into the optical fiber. The VCSEL was wire-bonded on the produced test board, and 10 mA of electricity was applied at 1.8 V using the power supply with an output power of 4 mW. The measured insertion loss is shown in Table 9.9. The average value and standard deviation of the insertion

TABLE 9.9 Measured Insertion Loss of the Integrated Microlens Array

Sample Number	Insertion Loss
1	−1.61 dB
2	−2.33 dB
3	−2.70 dB
4	−2.04 dB
5	−1.40 dB
6	−1.81 dB
7	−1.93 dB
8	−2.58 dB
9	−1.66 dB
10	−1.32 dB
11	−1.81 dB
12	−2.62 dB
13	−2.40 dB
14	−1.40 dB
15	−2.04 dB
16	−1.66 dB
17	−1.11 dB
18	−1.76 dB
19	−1.98 dB
20	−2.60 dB

loss were -1.94 and $0.47\,dB$, respectively. The measured insertion loss was less than $-3\,dB$, which was the design specification.

9.4 CONCLUSIONS

A separated microlens array to couple a VCSEL array to a multi-mode step-index optical fiber array was designed, fabricated, and tested. During the design stage, the design variables were optimized to maximize the coupling efficiency, and the effects of misalignments were analyzed. The final separated microlens array was replicated using a microcompression molding method with a metallic mold insert. The differences between the surface profile, surface roughness, and substrate thickness of the produced mold insert surface and the designed microlens surface were measured. Finally, the insertion loss of the separated microlens array was determined. The average value and standard deviation of the insertion loss were -1.49 and $0.74\,dB$, respectively. The measured insertion loss was less than $-3\,dB$, which was the design specification. Hence, a microlens array providing higher coupling efficiency between a multi-mode step-index optical fiber with a core diameter of $50\,\mu m$, and a VCSEL array was successfully produced using this technique.

Next, an integrated microlens array for coupling a VCSEL array to a single-mode step-index optical fiber array was designed, fabricated, and tested. During the design stage, the design variables were optimized to maximize the coupling efficiency, and the effects of misalignments were analyzed. The final integrated microlens array was replicated using a micro-UV-imprinting method. A UV-transparent mold was manufactured, and the differences between the actual and the designed mold surfaces were measured. Once the integrated array was replicated, the surface profile, surface roughness, and total height of the replications were measured, along with the insertion loss. The average value and standard deviation of the insertion loss were -1.94 and $0.47\,dB$, respectively. The measured insertion loss was less than $-3\,dB$, which was the design specification. This demonstrates that it is possible to obtain the desirable coupling efficiency when a microlens array with a small tolerance is produced directly on a VCSEL array. An integrated microlens array can provide higher coupling efficiency between a VCSEL array and a single-mode step-index optical fiber with a core diameter of $5\,\mu m$.

The results of this study demonstrate that an integrated microlens array fabricated using micro-UV-imprinting can be used to couple a multi-mode step-index optical fiber array and a VCSEL array. Because single-mode step-index optical fibers have smaller core diameters than multi-mode step-index optical fibers, it is generally more difficult to couple them with VCSEL arrays.

REFERENCES

1. S. R. Forrest, L. A. Coldren, S. C. Esener, D. B. Keck, F. J. Leonberger, G. R. Saxonhouse, and P. W. Shumate (1996) Optoelectronics in Japan and United States, final report. *Japanese Technology Evaluation Center (JTEC) panel report.*

2. K. Kurate, K. Yamauchi, H. Tanaka, H. Honmou, and S. Ishikawa (1999) A surface-mount single-mode laser module using passive alignment. *IEEE Transactions on Components, Packaging, and Manufacturing Technologies Part B: Advanced Packaging* 19, 554–559.

3. M. C. Cohen, M. F. Cina, E. Bassous, M. M. Opyrsko, J. L. Speidell, F. J. Frank, and M. J. Defranza (1992) Packaging of high-density fiber/laser modules using passive alignment techniques. *IEEE Transactions on Components, Hybrids, and Manufacturing Technology* 15, 944–954.

4. G. Nakagawa, K. Miura, S. Sasaki, and M. Yano (1996) Lens-coupled laser diode module integrated on silicon platform. *Journal of Lightwave Technology* 14, 1519–1523.

5. S. C. Wang, S. Chi, and W. H. Cheng (1998) A simple passive-alignment packaging technique for laser diode modules. *Materials Chemistry and Physics* 56, 189–192.

6. http://www.webster-dictionary.org/definition/vcsel.

7. S. Kang (2003) Fabrication Technology for Micro-Optics – Micromolding. *9th Microoptics Conference*, 106–109.

8. S. Moon, N. Lee, and S. Kang (2002) Fabrication of microlens array using microcompression molding with electroformed mold insert. *Journal of Micromechanics and Microengineering* 13, 98–103.

9. Z. D. Popovic, R. A. Sprague, and G. A. Neville Connell (1998) Technique for monolithic fabrication of microlens array. *Applied Optics* 27, 1281–1284.

10. D. L. MacFarlane, V. Narayan, J. A. Tatum, W. R. Cox, T. Chen, and D. J. Hayes (1994) Microjet fabrication of microlens arrays. *IEEE Photonics Technology Letters* 6, 1112–1114.

11. S. Kim, D. Kim, and S. Kang (2003). Replication of micro-optical components by ultraviolet-molding process. *Journal of Microlithography, Microfabrication, and Microsystems* 2, 356–359.

12. S. Moon, S. Ahn, S. Kang, D. Choi, and T. Je (2001) Fabrication of refractive and diffractive plastic micro-opitcal components using microcompression molding. *Device and Process Technologies for MEMS and Microelectronics II, Proceedings of SPIE* 4592, 140–147.

13. S. Moon, S. Kang, and J. Bu (2002) Fabrication of polymeric microlens of hemispherical shape using micromolding. *Optical Engineering* 41, 2267–2270.

14. J. M. Semior (1992) *Optical Fiber Communications - Principles and Practice*, Prentice Hall.

15. H. Kogelnik and T. Li (1996) Laser beams and resonators. *Applied Optics* 5, 1550–1567.

16. F. A. Rhaman, K. Takahashi, and C. H. Teik (2003) Theoretical analysis of coupling between laser diodes and conically lensed single-mode fibers utilizing ABCD matrix method. *Optics Communications* 215, 61–68.

17. Z. Tang, R. Zhang, S.K. Mondal, and F.G. Shi (2001) Optimization of fiber-optic coupling and alignment tolerance for coupling between a laser diode and a wedged single-mode fiber. *Optics Communications* 199, 95–101.

18. H. L. An (2000) Theoretical investigation on the effective coupling from laser diode to tapered lensed single-mode optical fiber. *Optics Communications* 181, 89–95.

19. J. Han, B. Min, and S. Kang (2008) Micro forming of glass microlens array using an imprinted and sintered tungsten carbide micro mold. International Journal of Modern Physics. B 22, 31&32, 6051–6056.

CHAPTER 10

Patterned Media

10.1 INTRODUCTION

Demand for ultra-high-density magnetic data storage devices has been increasing, and discrete track media (DTM) [1] and bit-patterned media (BPM) [2] have been proposed as solutions to overcome the limitations of conventional continuous magnetic media. Bit and grain sizes have been dramatically reduced to permit ultra-high storage density in magnetic storage media. As bit size decreases in conventional continuous magnetic media, medium noise between neighboring bits and the superparamagnetic effect increase inversely. DTM and BPM have physically separated magnetic domains, so they may be alternative solutions to overcome such difficulties.

For example, to ensure that information is stable over a period of 10 years, K_uV/k_BT (where K_u is the magnetic anisotropy, V is the magnetic switching volume, k_B is Botzmann's constant, and T is temperature) must remain greater than 40. However, the number of grains per bit must be maintained as areal density increases to ensure a sufficient signal-to-noise ratio (SNR). Thus, the grain size for magnetic stability must be reduced as areal density increases. Reduction of the grain size results in decreased magnetic switching volume (V).

DTM is separated physically and magnetically by grooves; the shape and width of the recording track is defined by the discrete track formed on the media surface. These discrete tracks and grooves can reduce the magnetic interference, referred to as 'cross-talk,' between adjacent tracks. In contrast to DTM, which is separated by discrete tracks along the radial direction on the disk substrate, BPM is separated by bit pattern in both the circumferential and radial directions. In BPM, the number of grains per bit can be reduced to 1. The magnetic switching volume is equal to the bit size, and the magnetic nanodot is thermally stable. Data can be stored in a bit as a single magnetic domain on the discrete nanoscale pattern of the media, as shown in

Micro/Nano Replication: Processes and Applications, First Edition. Shinill Kang.
© 2012 John Wiley & Sons, Inc. Published 2012 by John Wiley & Sons, Inc.

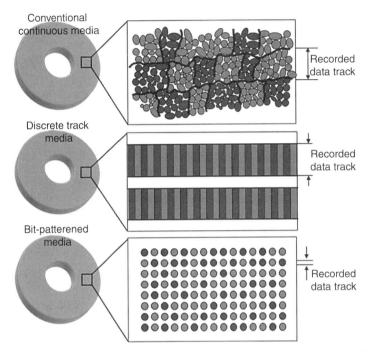

FIGURE 10.1   Schematic diagram of conventional continuous media, discrete track media (DTM), and bit-patterned media (BPM).

Figure 10.1. Nanoscale patterns for patterned media can be fabricated using various nanopatterning technologies such as electron beam (E-beam) lithography [3], laser interference lithography (LIL) [4], focused ion beam (FIB) [5], dry etching, etc. However, these procedures are unsuitable for mass production due to high cost and low throughput. Therefore, nano replication technology appears to be the most promising approach to low-cost fabrication of nanopatterns for patterned media [6]. One way to produce patterned media using nano replication is to deposit thin magnetic films on replicated pillar arrays [7–11]. This approach is superior to other techniques of patterning technology for patterned media; it allows any material to be deposited on the replicated nanopillars, and the process is easy and relatively low in cost. In addition, the technique permits large-area pattern generation.

Nanoimprinting lithography [7], which is categorized as a nano replication process, can produce nanoscale patterns at low cost. Nanoimprinting lithography requires two basic steps: imprinting and pattern transfer. During the mold step, a nanomold is pressed into a resist on a substrate to form nanopatterns. Then, reactive ion etching (RIE) is conducted to remove any residual resist in the compressed area during the pattern-transfer step. These processes are very advantageous when stack structures with fine alignment are required. However, a few technical problems must be overcome before nano imprinting technology can be used to mass-produce nanopatterns. For example, the surface quality of replicas can deteriorate as a result of interfacial phenomena such as sticking problems between the nanomold and the replica [8–10]. In addition, fabrication of transparent nanomolds for nanoimprinting is expensive [11].

Injection molding, which belongs to another class of nano replication processes, is another promising technology for the mass production of nano-/micropatterns with a relatively low aspect ratio. The technique is relatively low cost, and production cycles are short [12]. In addition, it does not involve an RIE process. However, injection molding has some technical problems that would need to be overcome to allow the mass production of high-aspect-ratio nanopatterns [13]. For example, polymer melt in the vicinity of mold surfaces solidifies rapidly during the filling stage of the injection molding process; this solidified layer greatly deteriorates replication quality [14]. Furthermore, a rigid nanomold that can endure harsh process conditions is required for nanoinjection molding applications [15]. Fabrication of a metallic mold with a nanopattern is a very difficult and expensive process compared to fabrication of a nonmetallic nanomold [16].

The following sections will introduce nanomold fabrication technology involving metallic and nonmetallic transparent materials for patterned media to achieve high storage densities of more than 1 Tbits/in^2.

First, a UV-imprinted polymeric nanomaster can be constructed from an original silicon master, followed by application of the nanoelectroforming process to fabricate the metallic nanomold [13]. Passive heating of the metallic nanomold can improve the replication quality of injection-molded nanopillars for patterned magnetic media [13]. Therefore, a passive heating system is usually designed and added to the injection molding procedure to delay solidification and increase the fluidity of the polymer melt [14]. This addition can yield the desired replication quality and height of the injection-molded nanopillars for patterned magnetic media.

10.2 FABRICATION OF A METALLIC NANO MOLD USING A UV-IMPRINTED POLYMERIC MASTER

E-beam lithography was used to fabricate the first master nanopattern in silicon [17]. Due to its high sensitivity and high resolution, ZEP520 resist was selected for use as the E-beam resist material during the E-beam lithography process. The E-beam lithography process was conducted using an EBPG-4HR system from LEICA. The silicon substrate was spin-coated with a positive ZEP520, with a resist thickness of 150 nm. The process used a beam current of 1 nA and an electron dose of 750 µC/cm^2. After E-beam exposure, the E-beam resist was developed in a 1:3 mixture of methylisobutylketone: isopropanol (MIBK:IPA) for 90 s. For fabrication of the silicon master, inductively coupled plasma (ICP) etching technology was used to etch the silicon substrate, using the E-beam master pattern as a barrier. The ICP etching process was conducted using a Multiplex ICP system from Surface Technology Systems (STS). The silicon etching process consisted of two steps. After the oxide layer on the E-beam master was removed through exposure to C$_{12}$ gas for 15 s in the initial etching step, the main etching step was implemented. The main etching process was carried out for 60 s using HBr$_2$ and O$_2$ with respective gas flow rates of 20 and 1. Chamber pressure was maintained at 2 mTorr throughout the etching process, and the chamber temperature was maintained at 20°C.

Analysis of the surface profiles of nanoholes on the E-beam master and the etched silicon master is very important to assess the fabrication process, but is also very difficult. In this case, surface profiles of nanohole patterns were investigated using a high-aspect-ratio atomic force microscopy (AFM) probe with nanosensors (SSS-NCHR); the trigonal pyramid-type probe of conventional AFM cantilevers is too blunt for insertion into deep nanoholes with sub-50 nm diameters. The high-aspect-ratio AFM probe has a half-cone angle at 2 µm of the high-aspect-ratio portion, typically below 5°. The guaranteed tip radius of curvature is less than 5 nm, so this probe is sharp enough to enter deep nanoholes with sub-50 nm diameters. Figure 10.2 presents scanning electron microscope (SEM) and AFM images of nanohole patterns on the E-beam master and the etched silicon master. The figure shows that the E-beam master and the etched silicon master are composed of uniform nanohole arrays with high density. Nanoholes on the etched silicon master were as small as 50 nm in diameter, 100 nm in pitch, and 35 nm in average depth.

Figure 10.3 presents SEM images of the 30 nm nanohole patterns on the etched silicon master. The figure shows that the E-beam master and the etched silicon master are composed of uniform nanohole arrays with high density. Nanoholes on the etched silicon master were as small as 30 nm in diameter and 50 nm in pitch. Therefore, nanopatterns on the E-beam master were successfully transferred to the silicon substrate using the ICP etching process. This fabricated silicon master can be used repeatedly as a master pattern in UV-imprinting processes.

Figure 10.4a presents the fabrication results for the silicon master with full tracks, with a track width of 1.3 mm and a track diameter of 40 mm. Figure 10.4b shows tracks of nanohole patterns with a diameter of 30 nm, a pitch of 50 nm, and a depth of 50 nm.

EBR and ICP etching process were used to fabricate the original silicon master with full tracks of nanohole patterns (diameter of 30 nm, pitch of

(a)

(b)

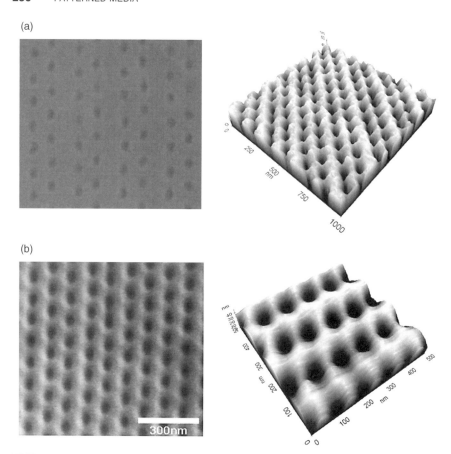

FIGURE 10.2 (a) SEM image and AFM image of nanohole arrays with a diameter of 50 nm, a pitch of 100 nm on E-beam nanomaster, (b) SEM image and AFM image of nanohole arrays with a diameter of 50 nm, a pitch of 95 nm on silicon master.

50 nm, track width of 1.3 mm, and track inner diameter of 40 mm). ZEP520 resist was spin-coated on the silicon substrate for the subsequent EBR process. After the coated substrate was set, it was moved to the designated position using a translation stage and then azimuthally rotated during the rotation stage during E-beam exposure. After the EBR process, the ICP etching process was implemented to etch the silicon substrate using the E-beam-lithographed master pattern as a barrier.

The DTM pattern can also be fabricated using the process detailed above. Some patterns are regular, whereas others are irregular and usually appear over the whole area of DTM. Figure 10.5 presents microscope and AFM images of DTM line patterns at a 70 nm pitch.

The polymeric nanomaster was replicated using the nano-UV-imprinting process with the silicon master. A self-assembled monolayer (SAM) of

FIGURE 10.3 SEM images of nanohole arrays with a diameter of 30 nm, a pitch of 50 nm (a) and line arrays with a width of 30 nm, a space of 50 nm (b) on silicon master.

fluoroctatrichlorosilane ($CF_3(CH_2)_8SiCl_3$) was deposited on the etched silicon nanomaster to improve the release properties of the silicon nanomaster prior to the nano-UV-imprinting process. The UV-mold system consisted of a UV lamp light source, an ellipsoidal reflector and collimating lens, a shutter to control duration of exposure, and a hydraulic pressure unit to apply pressure from the bottom.

The UV-imprinting process was carried out at room temperature with a curing dose of $20 \, mW/cm^2$ and a compression pressure of 90 kPa. To improve the flatness of the polymeric master, a glass substrate with high surface quality was prepared. First, a 1-methoxy-2-propanol-based adhesion

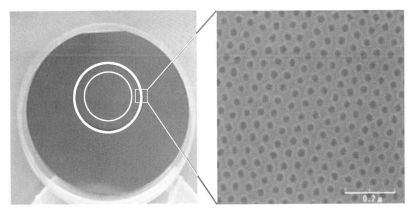

FIGURE 10.4 (a) Silicon master with full tracks having tracks width of 1.3 mm and track inner diameter of 40 mm; (b) SEM image of nanohole patterns having diameter of 30 nm and pitch of 50 nm on silicon master. Reprinted with permission from Ref. [17]. Copyright 2009, Institute of Electrical and Electronics Engineers.

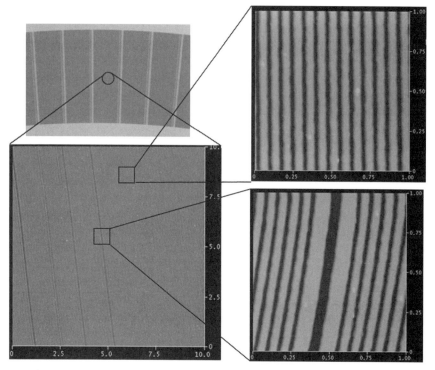

FIGURE 10.5   Microscope and AFM image of 70 nm pitch line pattern of DTM.

promoter was used to coat silanol radicals (-SiO) on the glass substrate. After the silicon master was pressed into the spin-coated photopolymer on the glass substrate, the photopolymer was exposed to UV light through the glass substrate. Nano-UV-imprinting has many advantages, including the ability to fill nanocavities, but it also has a problem: sticking between the polymer and the silicon master can degrade the replication quality and surface quality of imprinted nanopatterns. Here, silicon urethane acrylate UV-curable photopolymer, a photopolymer with low surface energy, was used as the UV-mold material to improve the releasing properties during the UV-imprinting process. Figure 10.6 shows SEM and AFM images of the nanopillar array on the polymeric master. The imprinted polymer pattern had an average height of about 35 nm, and the heights of the UV-imprinted polymeric pattern and the depth of the nanohole pattern on the silicon master deviated by less than 1 nm. The surface roughness (RMS) of the UV-imprinted pattern was 9.5 Å. These results indicate that the UV-imprinted polymeric master has a regular pillar pattern with good surface quality and that this polymeric master can be used as a cathode for electroforming processes during fabrication of nanomolds.

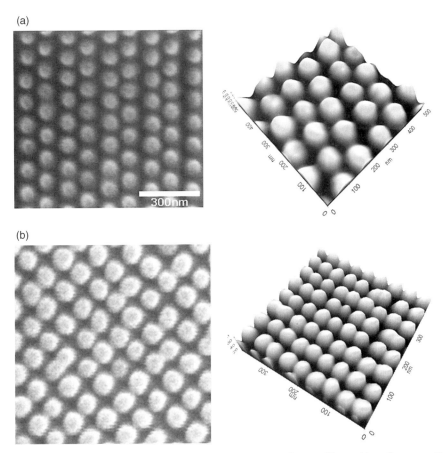

FIGURE 10.6 (a) SEM and AFM images of the array of nanopillars with a diameter of 50 nm, a pitch of 95 nm and a average height of 35 nm on a UV-imprinted polymeric master; (b) SEM and AFM images of the array of nanopillars with a diameter of 30 nm, a pitch of 50 nm, and a average height of 60 nm (aspect ratio = 2) on a UV-imprinted polymeric master.

The polymeric master with full tracks of nanopillar patterns was replicated on a glass substrate from the silicon master using the nano-UV-imprinting process.

The glass substrate must have high surface quality to guarantee the flatness of the polymeric master. To this end, an adhesion promoter was coated on the glass substrate using a spin-coating process at 3000 rpm for 25 s. Subsequently, it was baked on a hotplate at 110°C for 2 min. Fluorinated acrylate UV-curable photopolymer with a viscosity of 11.7 centipoise (cP) was dispensed on the glass substrate. Any air bubbles in the photopolymer dispensed on the glass substrate were removed in a vacuum chamber at 60 cm · Hg for 15 min.

Before initiating nano-UV-imprinting, the surface of the silicon master was rendered hydrophobic to ensure easy release; this entailed depositing a SAM of fluoroctatrichlorosilane (FOTS; $CF_3(CH_2)_8SiCl_3$) on the silicon master using plasma-enhanced chemical vapor deposition (PECVD). To enhance the adhesion of the SAM of fluoroctatrichlorosilane (FOTS) onto the silicon master, the silicon surface was treated with oxygen (O_2) plasma at 200 W for 120 s.

The nano-UV-imprinting process was carried out at room temperature with a curing dose of 20 mW/cm^2 and a compression pressure of 40 kgf/cm^2 for 30 s. The imprinted polymer patterns with pillar structure were then post-baked on a hotplate at 200°C for 30 min. Figure 10.6a presents replication results: the polymeric master had full tracks, with a track width of 1.3 mm and a track inner diameter of 40 mm. Figure 10.6b shows the uniform tracks of nanopillar patterns with a diameter of 30 nm and a pitch of 50 nm.

DTM patterns can also be replicated on the polymer substrate during the nano-UV-imprinting process, as shown in Figure 10.7. Figure 10.8 is a pattern image measured by AFM.

A nanomold was fabricated using seed layer deposition and a nanoelec-troforming process. The seed layer was the conduction layer in the electro-forming process and was deposited on the polymeric master through E-beam evaporation. The seed layer determines the surface quality and the mechanical properties of the nanomold because it functions as the mold surface after the electroforming process. A nickel-based material was selected for the seed layer due to its hardness, durability, and adhesion properties. During evaporation, evaporated metallic particles can attack the pillar pattern of the

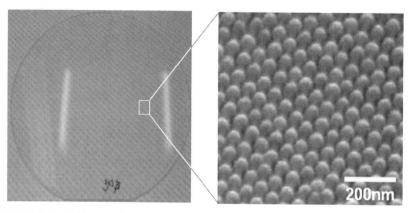

FIGURE 10.7   (a) Polymeric master with full tracks having track width of 1.3 mm, and track inner diameter of 40 mm. (b) SEM image of nanopillar patterns having diameter of 30 nm and pitch of 50 nm on polymeric master. Reprinted with permission from Ref. [17]. Copyright 2009, Institute of Electrical and Electronics Engineers.

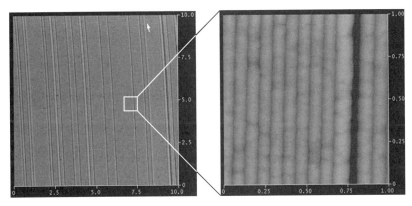

FIGURE 10.8 AFM image of polymer master surface for DTM.

polymeric master; this phenomenon can result in deteriorated surface quality of the mold surface. Therefore, deposition conditions were controlled to improve adhesion with the polymeric master and to prevent defects from forming on the polymeric master. Table 10.1 lists the optimal processing parameters for E-beam evaporation of the nickel seed layer.

After the seed layer was deposited on the polymeric nanomaster, an electroforming process using nickel sulfamate solution was carried out to fabricate the metallic nanomold. The electroforming process was controlled to suppress separation of the seed layer and development of residual stress in the electroformed layer. The electroforming bath (40 L capacity) was maintained at 44°C during the electroforming process by passing a current through it (density of 3.5–40 A/cm^2), and pH was maintained at 4.0. During the initial 60 min, a low current density was maintained (about 3.5 A/cm^2) to enable stable formation of the nickel-electroforming layer on the nickel seed layer. After the electroforming layer was complete, the current density was increased stepwise. The electroforming process was carried out very slowly to minimize residual stress and shrinkage. The electroforming process took 17 h to complete, and the resulting nickel-electroformed layer had a thickness of 295 μm.

After the electroforming process was complete, the polymeric nanomaster was removed from the electroformed layer, and the nickel seed layer adhered

TABLE 10.1 Optimized Processing Parameters for E-Beam Evaporation of the Nickel Seed-Layer on the Polymeric Master

Process Condition	Value
Temperature	45–50°C
Chamber pressure	6–10 Torr
Deposition rate	1 Å/s

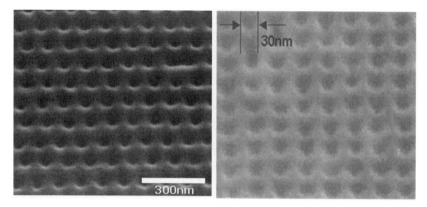

FIGURE 10.9 (a) SEM image of an array of nanoholes with a diameter of 50 nm, a pitch of 100 nm, and an average depth of 34.8 nm; (b) SEM image of an array of nanoholes with a diameter of 30 nm, a pitch of 50 nm on the metallic nanomold. The hardness of the nanomold was found to be 127.75 HRB.

to the nickel-electroforming layer. The fabricated mold had a diameter of 108 mm and a thickness of 0.3 mm. Figure 10.9a shows an SEM image of the nanomold pattern; nanoholes had a diameter of 50 nm and a pitch of 100 nm. Figure 10.9b shows an SEM image of the nanomold pattern; nanoholes had a diameter of 30 nm and a pitch of 50 nm. Therefore, the nanomold successfully transferred the nanoholes from the nanopillar pattern on the polymeric nanomaster. The nanomold was analyzed using an AKASHI measurement system (ATK-F1000) with a 100 Kgf tip, which revealed that the nanomold had hardness of 127.75 HRB. The nanomold would therefore be acceptable for use in nanoinjection molding processes, because conventional nickel molds used in nanoinjection molding processes have a hardness of about 126 HRB. The metallic mold had a surface roughness (RMS) of 4.9 Å, which guarantees a mirror-surface mold cavity. Therefore, this mold is acceptable for use in applications such as the replication of ultra-high-density data storages and optical communication.

Figure 10.10a presents the fabrication results using the metallic nanomold with full tracks; track width was 1.3 mm, and track inner diameter was 40 mm. Figure 10.10b shows tracks of nanohole patterns; these had a diameter of 30 nm, a pitch of 50 nm, and a depth of 30 nm. Therefore, full tracks of nanohole patterns on the metallic nanomold were successfully transferred from the full tracks of nanopillar patterns on the polymeric master.

After the nickel seed layer was deposited on the polymeric master, a nanoelectroforming process was carried out using a nickel sulfamate solution $(Ni(NH_2SO_3)_2 \cdot 4H_2O)$ to fabricate the metallic nanomold with full tracks of nanohole patterns. The electroforming process was controlled to suppress separation of the seed layer and development of residual stress in the

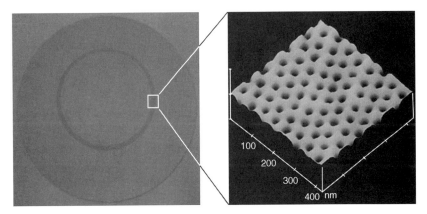

FIGURE 10.10   (a) Metallic nanomold with full tracks having track width of 1.3 mm and track inner diameter of 40 mm; (b) SEM image of nanohole patterns having diameter of 30 nm, pitch of 50 nm and depth of 30 nm on metallic nanomold. Reprinted with permission from Ref. [17]. Copyright 2009, Institute of Electrical and Electronics Engineers.

electroformed layer. The electroforming bath was maintained at 53°C, and pH was maintained at 3.84. During the ramping period of 450 s, current density was increased from 0–47.2 mA/cm^2 to ensure stable formation of the nickel-electroformed layer on the nickel seed layer. After the initial nanoelectroformed nickel layer was formed during the ramping period, current density was maintained at a constant value of 47.2 mA/cm^2. Deposition rate was precisely controlled at 12.2 nm/s to minimize residual stress and shrinkage in the electroformed layer. The nickel mold had a thickness of 300 μm after backpolishing and a diameter of 80 mm after wire cutting.

Figure 10.11 presents microscope and AFM images of the surface pattern on the fabricated nickel metal mold; counterclockwise, images are sized as follows: 5 × 5 μm, 1 × 1 μm, and 500 × 500 nm.

SAM is one candidate for use as an anti-adhesion layer in nano replication applications. SAM is physicochemically stable and can modify the surface properties of the mold without affecting the shape of the nanopatterns on the mold because monomolecules of 2–3 nm thickness are adsorbed on the mold surface. Many researchers have investigated SAM deposition on silicon or gold substrates, as SAM can be deposited on silicon or gold substrates after fairly simple pretreatments. However, silicon or gold substrates are not suitable materials for molds in nano replication intended for mass production. Nickel substrate is a superb material for molds intended for nano replication applications, but the oxide layer on the nickel surface deteriorates the anti-adhesion property of SAM-deposited nickel molds. Pretreatment to reduce this oxide layer on a nickel mold is difficult, which explains why most studies about SAM deposition on a nickel substrate have focused on the pretreatment process to reduce the oxide layer on the nickel surface.

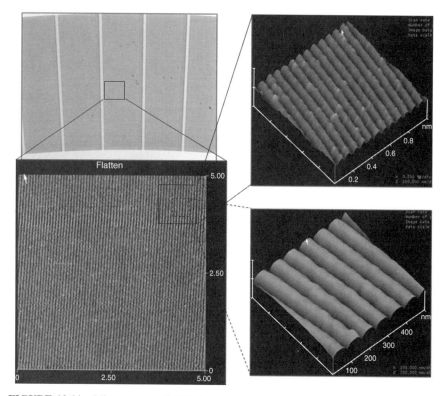

FIGURE 10.11 Microscope and AFM image of 70 nm pitch line pattern of metallic mold surface.

For example, Mekhalif et al. described how an electrochemical pretreatment can reduce the oxide layer on a nickel substrate and ensure SAM molecules are adsorbed on the nickel surface in an orderly fashion. *N*-dodecanethiol SAM on the nickel mold has a structure of 12 carbon chains, and its reaction group, -SH, links with nickel atoms. Then, van der Waals force ensures that molecular carbon chains, $-CH_2-$, keep close to other carbon chains, and these molecular combinations build an orderly monolayer on the nickel surface. In addition, the function group, $-CH_3$, decreases the monolayer's surface energy.

The present study is the first to apply *n*-dodecanethiol $(CH_3(CH_2)_{11}SH)$ SAM as an anti-adhesion layer to reduce the problem of adhesion between the nickel mold and replicated nanopatterns during nano replication processes. Before *n*-dodecanethiol SAM was deposited on the nickel mold, the surface of the nickel mold was pretreated to remove the oxide layer. Then, *n*-dodecanethiol SAM was deposited as an anti-adhesion layer on the nickel mold using the solution-deposition method. To examine the feasibility of *n*-dodecanethiol SAM as an anti-adhesion layer, contact angle and lateral friction force were measured at room temperature. To verify the feasibility of the SAM-deposited nickel mold, contact angle was measured at room temperature, and lateral friction force was measured at different normal forces from 4–12 nN.

First, the nickel mold was pretreated using the electrochemical reduction method. After this electrochemical pretreatment process, SAM was deposited on the nickel mold using the solution-deposition method. A n-dodecanethiol monolayer was formed by immersing the nickel substrates in a neat n-dodecanethiol solution.

As shown in Figure 10.12a, the contact angle between the nickel mold and D.I. water increased from 70.37°–109.22° after SAM deposition. Figure 10.12b shows the lateral friction force for the bare nickel mold and the SAM-deposited nickel mold at various normal forces. Lateral friction

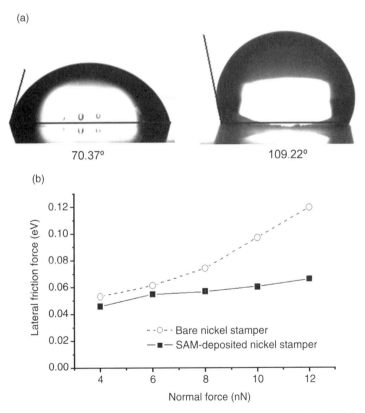

FIGURE 10.12   (a) Images of water contact angle of the bare nickel mold (70.37°) and the SAM-deposited nickel mold (109.22°) at room temperature. (b) Lateral friction force for different normal forces for the bare nickel mold and the SAM-deposited nickel mold. Reprinted with permission from Ref. [10]. Copyright 2007, American Institute of Physics.

forces at the various normal forces were 40% lower for the SAM-deposited nickel mold than for the bare nickel mold. As shown in Figure 10.12b, as normal force increased, SAM could markedly reduce the lateral friction force. This tendency of lateral friction forces at high normal forces indicates that a SAM-deposited nickel mold can effectively reduce the sticking force between a nickel mold and replicated parts at high molding pressures during the actual replication process. Results of lateral friction force and contact angle testing revealed that the uniform construction of monolayer molecules on the nickel mold and the -CH_3 group of the n-dodecanethiol SAM effectively reduce surface energy. Therefore, SAM appears to be suitable for use as an anti-adhesion layer for a nickel mold during the nano replication process.

10.3 FABRICATION OF PATTERNED MEDIA USING THE NANO REPLICATION PROCESS

As noted in the previous section, nanomold lithography requires two basic steps: imprinting and pattern transfer. During the imprinting step, a nanomold is pressed into resist on a substrate to form nanopatterns. Then, RIE is implemented to remove any residual resist in the compressed area during the pattern transfer step. Silicon or quartz molds are normally used as molds containing initial patterns in the nanomold lithography process due to their ease of fabrication [18]. However, silicon and quartz molds have short lifespans because they are too brittle to endure the process under cyclic pressing and thermal treatment. Polydimethylsiloxane (PDMS) or photo-polymers have also been used as molds for nanomold lithography [19], but polymer molds tend to be unsuitable for mass production due to their flexibility. Use of a metallic nanomold can prevent these problems, but would still require RIE, which is a very expensive process [20].

This section describes the application of nano replication technology to patterned media without an RIE process. Nanopatterns of holes and pillars were successfully transferred onto the polymeric substrate without using RIE, using the nickel nanomold and precise control of the nano replication process. The ferromagnetic islands deposited on the polymeric nanopatterns formed a single domain in their remanent state, demonstrating the applicability of the proposed methodology to patterned media.

A metallic nanomold was fabricated using electroforming, and then nanopatterned substrates were replicated using a nano replication process with the metallic nanomold.

Figure 10.13 provides an overview of the procedure. In the first step, the master nanopatterns were fabricated using E-beam lithography, and a nickel

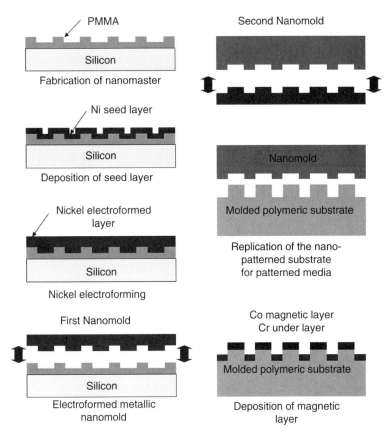

FIGURE 10.13 Fabrication procedures for patterned media using the nano replication technique. Reprinted with permission from Ref. [10]. Copyright 2007, American institute of physics.

seed layer was deposited on them. This was followed by the electroforming process to make the nickel nanomold. Next, the polymeric nanopatterned substrate was replicated using a nano replication technique. The surface profile and roughness of patterns on the polymeric substrate were measured and analyzed. Finally, the Cr underlayer and the Co magnetic layer were deposited on the nanopatterned polymeric substrate. Results confirmed that the nanopatterns were formed as a single magnetic domain state on the nanopatterned substrate. Each fabrication process is described in detail below.

Master nanopatterns were fabricated using E-beam lithography to replicate patterns at nanoscale sizes and pitches by a nano replication process. This process involved patterning a polymethyl methacrylate (PMMA) layer on a silicon substrate. The master nanopatterns were composed of arbitrary holes and pillars of various sizes. Figure 10.14 presents SEM and AFM images of

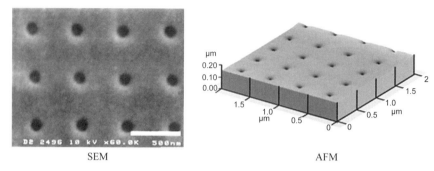

SEM   AFM

FIGURE 10.14 SEM and AFM images of the master nanopatterns with a diameter of 200 nm, a pitch of 500 nm and a depth of 100 nm. Reprinted with permission from Ref. [16]. Copyright 2004, Institute of Physics Science.

the master nanopatterns with a diameter of 200 nm, a pitch of 500 nm, and a depth of 100 nm.

Because the seed layer functions not only as a nanopattern on the nanomold but also as a conduction layer during electroforming, control of the deposition process and selection of the material for the seed layer are important. Here, the seed layer was deposited on the master patterns through E-beam evaporation to improve the deposition coverage and uniformity of metallic particles. A seed layer of nickel about 2000 Å was deposited because nickel can satisfy the requirements of sufficient hardness and thermal stability [21].

Process conditions must be controlled during E-beam evaporation to prevent separation of the seed layer and the electroformed layer from the E-beam-patterned layer. After deposition of the seed layer, the electroforming process was carried out using a commercial nickel sulfamate solution. This formed a nickel metallic nanomold with a thickness of 0.5 mm [9]. The electroforming process to suppress separation of the seed layer and the development of residual stress in the electroformed layer must be carefully controlled. For example, the deposition rate must be maintained at 1–2 μm/min[1], and the pH value must be maintained at 4.0–4.2. In addition, the electric current must be maintained at low values of 10–20 mA/cm^2 to slow the electroforming rate until the nickel metallic layer becomes stable.

After the electroforming process, the silicon wafer and E-beam resist can be removed, and the nanoscale pillars can be successfully transferred from the master nanopattern holes. Figure 10.15 shows SEM and AFM images of the metallic nanomold with a diameter of 200 nm, a pitch of 500 nm, and a depth of 100 nm.

A metallic nanomold was used for the molding process, which is considered a suitable mass-production process for replicating nanopatterns. The molding system can be modified from a microcompression molding

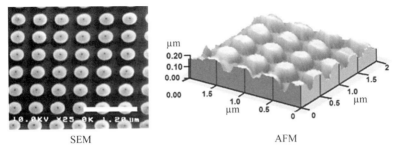

SEM   AFM

FIGURE 10.15   SEM and AFM images of the metallic nanomold with a diameter of 200 nm, a pitch of 500 nm and a depth of 100 nm. Reprinted with permission from Ref. [16]. Copyright 2004, Institute of Physics Science.

system [22]. PMMA powders are used for molding. The mold needs to be heated to a temperature above the PMMA glass transition temperature (T_g); as the mold temperature increases to temperatures greater than T_g, compression pressure is applied to replicate the patterns. This compression pressure is maintained while the mold is cooled.

When the demolding temperature has been reached, the replicated part can be released from the metallic mold. Figure 10.16 presents a schematic diagram of the molding system. Replication temperature and compressive pressure are the most dominant parameters governing the replication process; these determine the replication quality of replicated parts. If replication temperature and compressive pressure are too low, replicated patterns will not form because PMMA powders are not bonded and polymeric nano patterns will not fill a metallic nano mold. In contrast, if replication temperature and compressive pressure are excessively high, the resulting sticking effect can deteriorate replication quality. Therefore, controlling replication temperature and compressive pressure is important to improve the quality of replication in polymeric nanopatterns; the replication process is thus carried out at a temperature of around 210–220°C and a compressive pressure of 2–4 MPa. To evaluate the quality of replication in the replicated patterns, surface profiles of the polymeric hole patterns were analyzed using SEM and AFM images.

Figure 10.17 shows SEM and AFM images of the polymeric hole patterns with a diameter of 200 nm, a pitch of 500 nm, and a depth of 100 nm. Hole patterns had a surface roughness of ~8 Å. Table 10.2 summarizes typical surface roughness for the master nanopatterns, the metallic nanomold, and the polymeric nanopatterns. The SEM and AFM images confirmed that the replicated parts displayed the reverse polymeric patterns of the metallic nanomold. Further improvements to the surface quality of patterns appear to be possible; an anti-adhesion layer such as an SAM can reduce the sticking

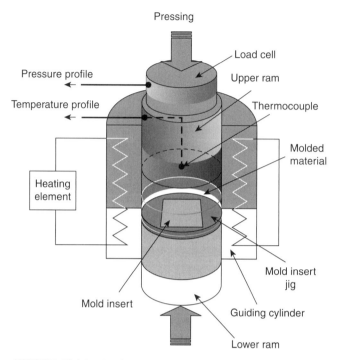

FIGURE 10.16 A schematic diagram of the replication system.

problem between mold and patterns. Applying an anti-adhesion SAM to the nickel mold can significantly improve the surface roughness of patterns [23].

To apply the proposed method to patterned media, a nanopatterned substrate with pillar patterns was replicated. This nanopatterned substrate had pillar patterns, some with a diameter of 200 nm and a pitch of 500 nm, and some with a diameter of 100 nm and a pitch of 250 nm.

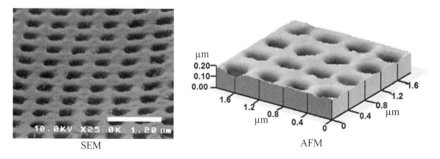

FIGURE 10.17 SEM and AFM images of the polymeric hole patterns with a diameter of 200 nm, a pitch of 500 nm and a depth of 100 nm. Reprinted with permission from Ref. [16]. Copyright 2004, Institute of Physics Science.

TABLE 10.2   Surface Roughness of the Master Nanopatterns, the Metallic Nanopatterns and the Polymeric Nanopatterns

	R_a (Å)
Master nanopatterns	3.29
Metallic nanomolds	6.42
Polymeric nanopatterns	7.55

Figures 10.18 and 10.19 show SEM and AFM images of the metallic nanomold and nanopatterned substrate for patterned media. A comparison of patterns on the replicated polymeric substrate revealed that the pillar patterns were not stable. In particular, some defects appeared on the substrate when the pillar diameter was 100 nm, and patterns were not uniformly distributed as designed. These results suggest that the replication of pillar patterns is more difficult than the replication of hole patterns, probably due to nonfilling and sticking problems. The nano replication process conditions will need to be further optimized, and the releasing method will need to be improved to solve those problems.

Finally, the magnetic layer was deposited onto the polymeric nanopatterns with diameters of 200 and 100 nm. First, a Cr underlayer (thickness of 100 Å) was deposited on the polymeric pillar patterns, followed by the Co magnetic

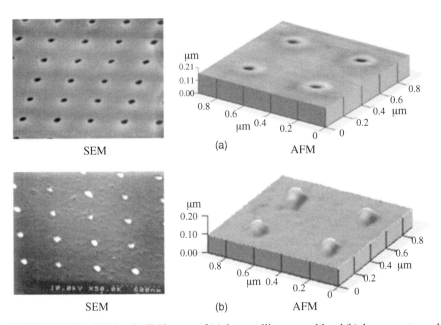

FIGURE 10.18   SEM and AFM images of (a) the metallic nanomold and (b) the nanopatterned substrate with a diameter of 200 nm and a pitch of 500 nm for patterned media. Reprinted with permission from Ref. [16]. Copyright 2004, Institute of Physics Science.

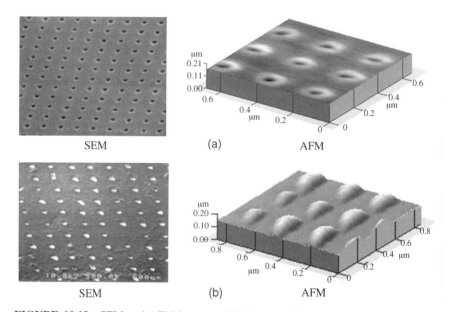

SEM   (a)   AFM

SEM   (b)   AFM

FIGURE 10.19 SEM and AFM images of (a) the metallic nanomold and (b) the nano-patterned substrate with a diameter of 100 nm and a pitch of 250 nm for patterned media. Reprinted with permission from Ref. [16]. Copyright 2004, Institute of Physics.

layer (thickness of 200 Å) in an ultra-high vacuum (UHV) sputtering system. To evaluate magnetic characteristics, the Co magnetic layer was saturated along the longitudinal direction of the patterns. The magnetic domain structure was analyzed using magnetic force microscopy (MFM). Figure 10.20 shows typical AFM and MFM images of the magnetic islands; some had a diameter of 200 nm and a pitch of 500 nm, and others had a diameter of 100 nm and a pitch of 250 nm. The figure illustrates that single magnetic-domain states were successfully generated on the nanopatterned substrate with a diameter of 200 nm. In contrast, pillars with a diameter of 100 nm exhibited multi-domain states, probably due to the nonuniform distribution of shapes and sizes of the nanopatterned magnetic pillars. The results indicate that a nano replication process using the metallic nanomold can be successfully applied to the mass production of patterned media.

10.4 FABRICATION OF PATTERNED MEDIA USING INJECTION MOLDING

In recent years, nontraditional technologies for the nanofabrication of ultra-high density perpendicular patterned magnetic media have received much attention. One approach to producing patterned media is to deposit thin

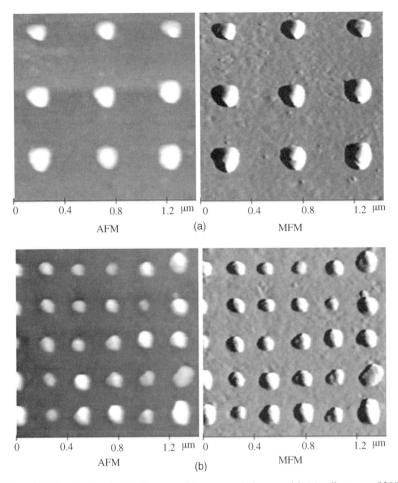

FIGURE 10.20   AFM and MFM images of the magnetic layers with (a) a diameter of 200 nm and a pitch of 500 nm and (b) a diameter of 100 nm and a pitch of 250 nm for patterned media. Reprinted with permission from Ref. [16]. Copyright 2004, Institute of Physics.

magnetic films on pre-etched pillar arrays prepared with nano replication technology [19,24]. However, a few technical problems must be overcome before nano replication technology can be used in the mass production of patterned magnetic media. For example, interfacial phenomena such as sticking between the nanomold and the replica surface can deteriorate the quality of replicas [10,25]. In addition, cleaning processes are required to remove contamination from the mold at each replication step. Injection molding is another promising technology for the mass production of nano-/ micropatterns with a relatively low aspect ratio because it is inexpensive and has a shorter cycle time replication process [15,26]. In addition, injection molding does not involve the RIE process.

This section discusses the feasibility of injection molding technology to make patterned media, e.g., small form factors or nonrotating types [15,27]. In particular, it will focus on two important technical issues that must be overcome before injection molding technology is used for the mass production of patterned magnetic media. The first issue is related to fabrication of a metallic nanomold, which is necessary for injection molding. Generally, metallic nanomolds with nanocavity structures are fabricated through electroforming processes using a photoresist (PR) patterned master or silicon master, but this method is not cost-effective because the expensive original master cannot be reused after the mold is fabricated [15,16]. Another option is to obtain a UV-imprinted polymeric nanomaster from the original silicon nanomaster and then use a nanoelectroforming process to fabricate the metallic nanomold. In this case, the expensive silicon master mold can be reused many times as a master mold in nano-UV imprinting, because the UV-imprinting process is performed at low pressures and room temperature [25]. The second issue is related to passive heating of the metallic nanomold to increase the replication quality of the injection-molded nanopillars for patterned magnetic media. In the conventional injection molding process, the polymer melt in the vicinity of the mold surfaces rapidly solidifies during the filling stage, and the solidified layer greatly deteriorates the quality of replication [15,27]. In the proposed injection molding procedure, a passive heating system has been designed and constructed to delay this solidification and increase the fluidity of the polymer melt. This can improve replication quality and yield the desired height of injection-molded nanopillars for patterned magnetic media.

Finally, a magnetic Co–Cr–Pt (CCP) alloy film is deposited on the replicated nanopillars to convert the polymeric nanopillars into a patterned magnetic media. The magnetic domain states and recordability of the magnetic pillars were assessed using MFM measurement. The effect of replicated pattern height on the readback signal characteristics of magnetic nanopillar tops and trench bottoms was also analyzed [14].

Figure 10.21 is a schematic diagram of the cost-effective technology currently used to fabricate metallic nanomolds and how it is used in injection molding to make nanopillars for patterned media. Section .1 explains how the first nanopatterns are fabricated on the silicon wafer using E-beam patterning with ZEP520 resist and ICP etching. The polymeric nanomaster is then replicated using UV imprinting with the etched silicon nanomaster. Before the nano-UV-imprinting process, SAM of fluoroctatrichlorosilane [CF3(CH2)8SiCl3] can be deposited on an etched silicon nanomaster to improve its releasing property [24]. Next, a nickel seed layer is deposited on the UV-imprinted polymeric nanomaster using E-beam evaporation, and a metallic nanomold is fabricated using a nanoelectroforming process with a

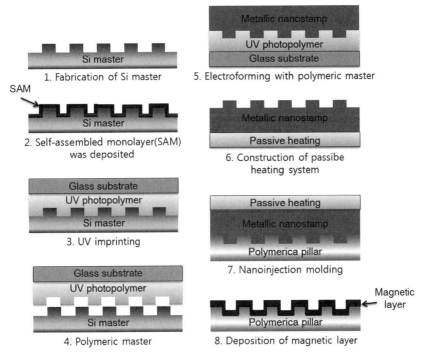

1. Fabrication of Si master

SAM

2. Self-assembled monolayer(SAM) was deposited

3. UV imprinting

4. Polymeric master

5. Electroforming with polymeric master

6. Construction of passibe heating system

7. Nanoinjection molding

8. Deposition of magnetic layer

FIGURE 10.21 Schematic diagram of our novel cost-effective technology for the fabrication of metallic nanomolds by using a UV-imprinted polymeric nanomaster, and its application to injection molding of nanopillars for patterned media with a passive heating system. In this process, a UV-imprinted polymeric pattern with excellent chemical resistance and stability is used instead of an E-beam master as a master in the nanoelectroforming process. Reprinted with permission from Ref. [13]. Copyright 2008, APEX/JJAP.

nickel sulfamate solution [$Ni(NH_2SO_3)_24H_2O$]. Figures 10.22a–c show SEM and AFM images of nanocavities and nanopillars on the silicon nanomaster, the polymeric nanomaster, and the metallic nanomold, respectively. The average height of nanopillars on the polymeric nanomaster was 35 ± 1 nm. The average difference in depth between the silicon nanomaster and the polymeric nanomaster was less than 2 nm. The metallic mold had an RMS surface roughness value of 4.9 Å, so it can be considered to have a mirror surface.

Polycarbonate (Lexan OQ 1020c-112) was used as the polymer resin in the injection molding. Nanopillar patterns with good surface quality can be successfully replicated on replicated substrates using the injection-replication method and passive heating, which can be ensured by using a polyimide thermal insulation layer with a thickness of 75 mm.

To make patterned media from injection-molded nanopillars, the magnetic layer is deposited on the polymeric pillar patterns using a high vacuum (HV) sputtering system. A CCP magnetic alloy is deposited to a thickness of 12 nm on

FIGURE 10.22 SEM and AFM images of nanocavities and nanopillars on a (a) silicon nanomaster, (b) polymeric nanomaster, and (c) metallic nanomold. The nanocavities and nanopillars are as small as 50 nm in diameter, 100 nm in pitch. (d) SEM and AFM image of the injection-molded nanopillars obtained using the passive heating injection mold; the average height of the molded pattern is 34.4 nm. Reprinted with permission from Ref. [13]. Copyright 2008, APEX/JJAP.

the Ru/Ta underlayer. The CCP layer exhibits a static coercive force (H_c) of 3400 Oe, which is sufficiently high for use in perpendicular patterned magnetic media. The Fourier component of the readback voltage can be calculated to estimate the readback signal characteristics of magnetic films on pillar tops and trench bottoms. The normalized readback stray signal from the trench bottom decays exponentially with the height to pitch ratio, h/p; it can expressed as follows: where h and p are the height and pitch of the nanopillars. Thus, nanopillars with a high aspect ratio are essential to obtain a high SNR in magnetic recording. Experimental results revealed that the injection-molded nanopillars without passive heating had a height of 12.8 nm, and injection-molded pillars with passive heating had a height of 34.4 nm, so passive heating appears to reduce magnetic noise from the trench bottoms by approximately 50.6%. This suggests that injection-molded nanopillars formed using passive heating can achieve a high SNR in readback signals from perpendicular-patterned magnetic media.

MFM was conducted to assess the magnetic domain structures of the magnetic films on injection-molded pillars; this involved scanning and measuring the magnetic force between the MFM tip and a magnetized sample. Figure 10.23 shows the surface structure and magnetic domain structures of magnetic nanopillars, which were produced using the proposed injection molding technique with passive heating before and after magnetic saturation. Figure 10.23a shows an SEM image of the surface structure of the media. The magnetic domain image shown in Figure 10.23b is the as-deposited film, which was expected to be in a demagnetized state.

The white and black circles in the MFM image in Figure 10.23b indicate that each pillar was actually randomly magnetized in an upward (white) or

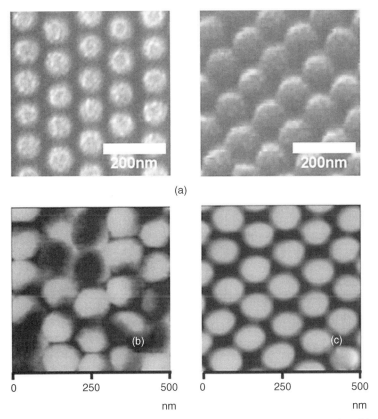

FIGURE 10.23 Surface and magnetic images for magnetic nanopillars produced by using our injection molding with passive heating. (a) SEM images of the media surface, (b) MFM image for the as deposited film in the demagnetized state (pillars are randomly magnetized but their magnetization is either "upward" (white) or "downward" (black)), and (c) MFM image at the remanent state after magnetic saturation (the white on the pillars indicates that they are saturated in only one direction). Reprinted with permission from Ref. [13]. Copyright 2008, APEX/JJAP.

downward (black) direction. Pillars with only black or white circles were uniformly magnetized. In other words, each pattern island is a single magnetic domain. To test the recordability of magnetic islands on the media, a magnetic field of more than 15 kOe was applied to saturate all magnetic islands in one direction. As shown in Figure 10.23c, all the single domains were switched and magnetically saturated along the applied field direction.

Next, the magneto-optic kerr effect (MOKE) was used to evaluate magnetic properties. As shown in Figure 10.24b, the magnetic film with a thickness of 30 nm had a coercive force (H_c) about 3400 Oe higher than the magnetic film with a thickness of 23 nm. These results indicate that the 30 nm thick CCP alloy on polymeric nanopillar arrays had a coercive force sufficiently high to warrant using the proposed method for perpendicular magnetic-patterned media. Together, these observations confirm that the molded nanopillars produced using injection molding with passive heating are suitable for the production of patterned magnetic media.

10.5   MEASUREMENT AND ANALYSIS OF MAGNETIC DOMAINS OF PATTERNED MEDIA BY MAGNETIC FORCE MICROSCOPY

With the increasing demand for patterned media, many researchers have focused on measuring and analyzing high-density patterned media. Among the various measurement technologies for patterned magnetic media, MFM is a very promising technology for imaging magnetic domains of magnetic nanopatterns. The process involves relatively simple sample preparation and measurement, and high-resolution images can be obtained under normal laboratory conditions. In addition, MFM measurement can be carried out using a conventional scanning probe microscope (SPM) system.

Certain technical problems arise when measuring high-density patterned media with sub-50 nm pillars. For example, as the features of nanopillars on the patterned media become smaller, higher-resolution MFM tips are needed to measure the smaller magnetic domain. Furthermore, as the features of nanopillars on the patterned media become smaller, properties of the MFM tip and tip–sample distance can seriously influence the tip–sample interface during MFM measurement. This section focuses on some of the technical issues of MFM measurement, particularly the geometry of the MFM tip and the lift height [29].

The optimum lift height and the distance between the MFM tip and the patterned substrate must be determined to obtain a high-resolution magnetic domain image of sub-50 nm patterned media. The present study investigated the relationship between MFM signal and lift height both numerically and experimentally to obtain high-resolution magnetic domain imaging.

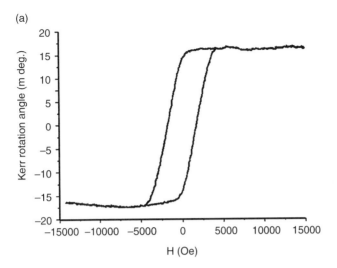

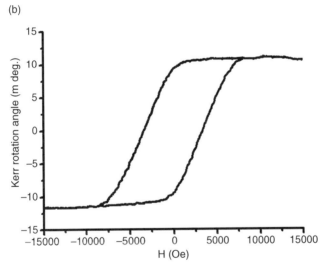

FIGURE 10.24   A perpendicular MOKE hysteresis loop of the Co–Cr–Pt layer on the Ru/Ta underlayer; (a) first type magnetic film with the thickness of 23 nm; (b) Second type magnetic film with the thickness of 30 nm. Reprinted with permission from Ref. [17]. Copyright 2009, Institute of Electrical and Electronics Engineers.

Generally, the magnetic force acting on a tip can be obtained by integrating the tip–sample force density as follows:

$$F = \mu_0 \nabla (m \cdot H) \tag{10.1}$$

where m is the magnetic moment of the MFM tip, H is the magnetic stray field from the sample, and μ_0 is the vacuum permeability. Thus, MFM can measure quantitative information about the magnetic stray field from a sample.

Based on Eq. (10.1) in section 10.3, as the lift height of the MFM tip increases, the magnetic stray field and the magnetic signal from the media can decay exponentially as a function of the lift height (h_{lift}). The relationship between MFM magnetic signal (F) and lift height (h_{lift}) can be expressed as follows:

$$F \propto e^{-kh_{lift}} = e^{-\pi h_{lift}/p} \tag{10.2}$$

Thus, a smaller lift height in MFM measurement is necessary to obtain a high-resolution MFM scanning image. However, when lift height becomes excessively small, atomic force from the surface structure can result in noise during MFM scanning.

Based on the numerical analysis, the relationship between the MFM signal and lift height was investigated experimentally. A multi-mode SPM system (Veeco Co.) was used as the MFM measurement system, and a cobalt-coated MFM cantilever (Nanosensors Co.) was used as the tip. The MFM tip radius was 25 nm, and the tip had a film thickness of 20 nm. The MFM tip had a magnetic coercivity (H_c) of 0.75 Oe and a remnant magnetic moment of about 225 emu/cm^3.

Within the SPM system, the MFM mode was measured together with the AFM by applying the lift mode. The lift mode consisted of two pass scanning processes: atomic force for the short-range force and magnetic force for the long-range force. Figure 10.25 shows how topographic scanning was obtained during the first pass by measuring the atomic force. During the second pass, atomic force was eliminated from the MFM signal, and magnetic information was recorded by lifting the MFM tip from the sample. Lift height during MFM measurement generally ranged from 10–200 nm.

The relationship between the MFM signal and the lift height was then evaluated; this involved using the same patterned media substrate with sub-50 nm patterns to evaluate the MFM of patterned media at various lift heights. The patterned media substrate sub-50 nm patterns were fabricated using the injection molding process with passive heating. A magnetic field of more than 15 kOe was applied to saturate all magnetic islands on the patterned media in one direction. The MFM tip was magnetized using a hard magnetic bar before each experiment.

The same patterned media substrate was scanned using the same MFM tip at various lift heights from 10–120 nm, the typical lift-height range for MFM measurement. Lift heights were set to 10, 30, 50, 70, 90, 110, and 130 nm to observe the effect of lift height on the magnetic signal of the same patterned media substrate.

Figure 10.26 presents MFM images of 50 nm patterned media substrate at various lift heights (h_{lift}). As shown in Figure 10.20, as the lift height decreased

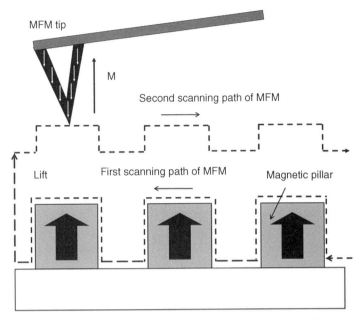

FIGURE 10.25 Schematic diagram of MFM measurement system of patterns media. Reprinted with permission from Ref. [17]. Copyright 2009, Institute of Electrical and Electronics Engineers.

but remained over 35 nm, resolution of the MFM images improved. When the lift height was over 35 nm, the resolution of the MFM images tended to be inversely related to lift height, and as the lift height decreased, the frequency of the magnetic signal slowly increased. However, when the lift height was below 35 nm, MFM images of the patterned media were distorted compared to AFM images.

Figure 10.27 presents the relationship between the magnetic signal and various lift heights ($h_{lift} = 30$–130 nm). As shown in Figures 10.26 and 10.27, when lift height was around 30 nm (approximately the pattern height), the frequency of the magnetic signal value increased markedly, and the resolution of MFM images was distorted. These results suggest that atomic force from the surface structure acts as noise during the MFM scanning process due to the excessive proximity of the tip.

Experimental analyses of MFM resolution and lift height supported the results of the numerical analysis. The readback voltage of a magnetic medium, V_x, can be expressed as follows [14]:

$$V_x(k) = KM_r e^{-k(a+d)} \frac{1 - e^{-k\delta}}{k} \qquad (10.3)$$

(a) h_{lift} = 70 nm (b) h_{lift} = 90 nm (c) h_{lift} = 110 nm

(d) h_{lift} = 35 nm (e) h_{lift} = 40 nm (f) h_{lift} = 50 nm

(g) h_{lift} = 30 nm (h) h_{lift} = 31 nm (i) h_{lift} = 32.5 nm

FIGURE 10.26 MFM images of the same patterned-media substrate with a various lift height (30–110 nm). Reprinted with permission from Ref. [17]. Copyright 2009, Institute of Electrical and Electronics Engineers.

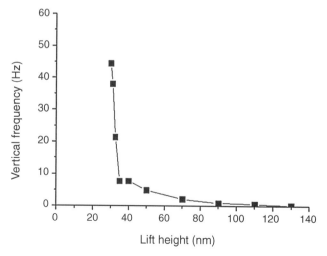

FIGURE 10.27 Relationship of magnetic signal and a various lift height (h_{lift} = 30–130 nm).

where K is a constant. Thus, readback signal characteristics of patterned media can be affected by M_r, δ, k, a, and d, which are the remanence magnetization of the medium (magnetic film), the medium (magnetic film) thickness, the wave vector, transition parameter, and magnetic spacing, respectively.

Finally, qualitative and quantitative analysis of MFM revealed that the optimum lift height of the MFM tip for high-resolution scanning images is 35 nm.

10.6 CONCLUSIONS

Master nanopatterns containing nanoscale holes and pillars were fabricated using E-beam lithography. Electroforming was used to make the nickel nanomold, and then the nanoimprinting or injection molding was performed to replicate the polymeric nanopatterns without an additional etching process.

This chapter investigated two important technical issues that must be overcome before nanoimprinting and injection molding technology can be applied in the mass production of patterned magnetic media. First, metallic nanomolds were successfully fabricated using a UV-imprinted polymeric nanomaster. Second, replication quality and pattern height of injection-molded nanopillars were improved by using a passive heating system, including a thick insulation layer. In addition, magnetic noise from the trench bottoms of replicated patterns was reduced by achieving a high aspect ratio, which can be obtained through appropriate techniques such as passive heating. Each individual magnetic nanopillars formed using nanoimprinting and injection molding with passive heating acts as a single magnetic domain state.

The chapter also investigated the relationship between MFM signal and lift height, both numerically and experimentally. As the lift height of the MFM tip decreased, the resolution of MFM images improved. However, when the lift height became excessively small, the atomic force from the surface structure could act as noise during the MFM scanning process. Finally, analyses revealed that the optimum lift height of the MFM tip for high-resolution scanning image is 35 nm.

These results confirm that the proposed methodology is appropriate for use in the fabrication of perpendicular patterned magnetic media. Further research will be required to reduce the pattern size and test the media performance.

REFERENCES

1. K. Hattori, K. Ito, Y. Soeno, M. Takai, and M. Matsuzaki (2004) Fabrication of discrete track perpendicular media for high recording density. *IEEE Transactions on Magnetics* 40, 2510–2515.

2. R. L. White, R. M. H. New, and R. F. W. Pease (1997) Patterned media: a viable route to 50 Gbit/in^2 and up for magnetic recording. *IEEE Transactions on Magnetics* 33, 990–995.

3. T. Aoyama, K. Uchiyama, Y. Kagotani, K. Hattori, Y. Wada, S. Okawa, H. Hatate, H. Nishio, and Sato (2000) Fabrication and properties of Co–Pt patterned media with perpendicular magnetic anisotropy. *IEEE Transactions on Magnetics* 37, 1646.

4. X. Hao, M. Walsh, M. Farhoud, C. A. Ross, H. I. Smith, J. Q. Wang, and L. Malkinski (2001) In-plane anisotropy in arrays of magnetic ellipses. *IEEE Transactions on Magnetics* 36, 2996–2998.

5. J. Lohau, A. Moser, C. T. Rettner, M. E. Best, and B. D. Terris (2001) Writing and reading perpendicular magnetic recording media patterned by a focused ion beam. *Applied Physics Letters* 78, 990–992.

6. S. Y. Chou and P. R. Krauss (1997) Mold lithography with sub-10 nm feature size and high throughput. *Microelectronic Engineering* 35, 1–4 237–240.

7. S. Y. Chou, P. R. Krauss, and L. Kong (1996) Nanolithgraphically defined magnetic structures and quantum magnetic disk. *Journal of Applied Physics* 79 (8), 6101–6106.

8. N. Lee, Y. Kim, and S. Kang (2004) Temperature dependence of Anti-adhesion between the stamper with sub-micron patterns and the polymer in nano molding processes. *Journal of Physics D: Applied Physics* 37(12) 1624–1629.

9. N. Lee, S. Moon, S. Kang, and S. Ahn (2003) The effect of wettability of nickel mold insert on the surface quality of molded microlenses. *Optical Review* 10(4), 290–294.

10. N. Lee, S. Choi, and S. Kang (2006) Self-assembled monolayer as an anti-adhesion layer on a nickel nanostamper in the nano replication process for optoelectronic applications. *Applied Physics Letters* 88(7) 073101.

11. I. Maximov, E. L. Sarwe, M. Beck, K. Deppert, M. Graczyk, M.H. Magnusson, and L. Montelius (2002) Fabrication of Si-based nanomold stamps with sub-20 nm features. *Microelectronic Engineering* 61–62, 449–454.

12. A. W. McFarland, M. A. Poggi, L. A. Bottomley, and J. S. Colton (2004) Injection molding of high aspect ratio micron-scale thickness polymeric micro-cantilevers. *Nanotechnology* 15, 1628–1632.

13. N. Lee, J. Han, J. Lim, M. Choi, Y. Han, J. Hong, and S. Kang (2008) Injection molding of nanopillars for perpendicular patterned magnetic media with metallic nanostamp. *Japanese Journal of Applied Physics* 47, 1803–1805.

14. M. J. Liou and N. P. Suh (1989) Reducing residual stresses in molded parts. *Polymer Engineering and Science* 29, 441–447.

15. T. G. Bifano, H. E. Fawcett, and P. A. Bierden (1997) Precision manufacture of optical disc master stampers. *Precis Engineering* 20, 53.

16. N. Lee, Y. Kim, S. Kang, and J. Hong (2004) Fabrication of metallic nanostamper and replication of nanopatterned substrate for patterned media. *Nanotechnology* 15(8), 901–906.

17. H. Kim, S. Shin, J. Han, J. Han, and S. Kang (2009) Fabrication of metallic nano stamp to replicate patterned substrate using electron-beam recording, nanoimprinting and electroforming. *IEEE Transactions on Magnetics* 45(5), 2304–2307.

18. Y. Hirai, N. Takagi, H. Toyota, S. Harada, T. Yotsuta, and Y. Tanaka (2001) Nanochamber fabrication on an acrylic plate by direct nanomold lithography using quartz mold. *International Conference on Microprocesses and Nanotechnology (Shimane, Japan)*, 104–105.

19. G. M. McClelland, M. W. Hart, C. T. Rettner, M. E. Best, K. R. Cartner, and B. D. Terris (2002) Nanoscale patterning of magnetic islands by imprint lithography using a flexible mold. *Applied Physics Letters* 81, 1483–1485.

20. Y. Hirai, S. Harada, S. Isaka, KobayashiM, and Y. Tanaka (2002) Nanomold lithography using replicated mold by Ni electroforming. *Japanese Journal of Applied Physics* 41, 4186–4189.

21. S. Moon, N. Lee, and S. Kang (2003) Fabrication of a microlens array using mircocompression molding with an electroformed mold insert. *Journal of Micromechanics and Microengineering* 13, 98–103.

22. S. Moon, S. Kang, and J. Bu (2002) Fabrication of polymeric microlens of hemispherical shape using micro-molding. *Optical Engineering* 41, 2267–2270.

23. S. Choi, N. Lee, Y. Kim, and S. Kang (2004) Properties of self-assembled monolayer as an anti-adhesion layer on metallic nanostamper. *The Nanotechnology Conference (Boston)* 3, 350–353.

24. J. Moritz, S. Landis, J. C. Toussaint, P. Bayle-Guillemaud, B. Rodmacq, G. Casali, A. Lebib, Y. Chen, J. P. Nozieres, and B. Dieny (2002) Patterned media made from pre-etched wafers: A promising route toward ultrahigh-density magnetic recording. *IEEE Transactions on Magnetics* 38, 1731.

25. N. Gadegaard, S. Mosler, and N. B. Larsen (2003) Biomimetic polymer nanostructures by injection molding. *Macromolecular Materials and Engineering* 288, 76.

26. L. J. Guo (2004) Topical review: recent progress in nanoimprint technology and its applications. *Journal of Physics D: Applied Physics* 37(11), R123–R141.

27. A. W. McFarland, M. A. Poggi, L. A. Bottomley, and J. S. Colton (2004) Injection moulding of high aspect ratio micron-scale thickness polymeric microcantilevers. *Nanotechnology* 15, 1628.

28. Y. Kim, J. Bae, H. Kim, and S. Kang (2004) Modelling of passive heating for replication of sub-micron patterns in optical disk substrates. *Journal of Physics D: Applied Physics* 37, 1319.

29. S. X. Wang and A. M. Taratorin (1998) *Magnetic Information Storage Technology*, Academic Press, London, pp. 147.

30. M. Choi, J. M Yang, J. Lim, N. Lee, and S. Kang (2009) Measurement and analysis of magnetic domain properties of high-density patterned media by magnetic force microscopy. *IEEE Transactions on Magnetics* 45(5), 2308–2311.

CHAPTER 11

Optical Disk Drive (ODD)

11.1 INTRODUCTION

There is increasing demand for high-performance, low-cost, and large-capacity storage, brought about by the growing popularity of applications such as HDTV and digital broadcasting. To meet these requirements, a variety of information storage systems, including magneto-optic (MO) disk drives [1], phase-change optical disk drives [2], near-field recording (NFR) storage [3], Super-RENS [4], holographic optical storage [5], and two-photon storage [6] have been developed, with many of these systems being commercially available.

Among these information storage systems, optical storage devices, including compact-disk read-only memory (CD-ROM), digital versatile disk (DVD), and high-density digital versatile disk (HD-DVD) have become widely used and are commercially available products largely, because these optical data storage devices are portable, low-cost, and have good compatibility standards.

With the ever-increasing demand for high capacity in optical data storage, there is a need to develop new technologies. One such development is to make double-layered or multi-layered optical disks and another is to reduce the size of the features on the disk. These two solutions have to be performed within the limits of the existing disk format, since the standardization of optical disks restricts the form factor for reasons that include portability, distribution issues, and compatibility with the existing format.

Reducing the feature size on optical disks means that the areal density, namely the capacity per unit area, increases. The track pitch is expected to be narrower in future formats than the present DVD standard of 0.74 μm, which will require the size of the laser spot to be reduced. To do so, the wavelength, λ, should be decreased and the numerical aperture (NA) of the lens should be increased. The wavelength of the lasers used in existing optical storage

Micro/Nano Replication: Processes and Applications, First Edition. Shinill Kang.
© 2012 John Wiley & Sons, Inc. Published 2012 by John Wiley & Sons, Inc.

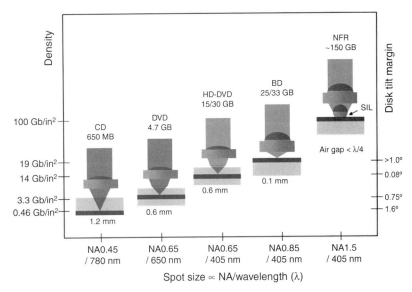

FIGURE 11.1 Technology trend of spot size and disk tilt margin for high-density optical data storage. Spot size is determined by numerical aperture (NA) and wavelength of laser source and spot size and thin cover-layer affect the areal density associated with optical storage capacity.

devices has shortened as new standards have been developed. For example, the CD-ROM format uses 780 nm, whereas a DVD uses 650 nm. The NA of the lenses, including the objective and immersion lenses, can also be increased to help to reduce the laser spot size. Both a solid immersion lens (SIL) [7] and a liquid immersion lens [8] can increase the NA. These techniques have been adopted for near-field recording storage, and research for next-generation formats is ongoing.

Figure 11.1 shows the trend of increasing capacity in optical data storage. As the NA increases and the wavelength of the light source decreases, higher-density optical data storage can be achieved. However, the disk-tilt margin worsens, because the optical disk substrate and cover layer both become thinner.

Figure 11.2 shows a schematic diagram of a near-field optical data storage system. It includes the optical disk, optical pickup, including the objective lens, SIL and actuator, two lasers (having different wavelengths), and a photo detector. As the wavelength becomes shorter and the NA increases, the laser spot size on the cover layer can be reduced but, aberration generally tends to increase. Auxiliary components, such as microcompensatory lenses [9,10], can be used to compensate for such aberration.

Recently, high-density DVDs (HD-DVD) [12] and optical disks for digital video recording (DVR) using blue lasers have been developed. HD-DVD and DVR differ from the conventional DVD system in that the HD-DVD or DVR

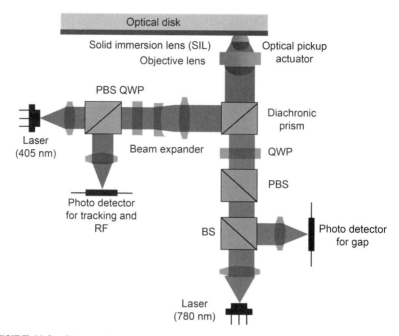

FIGURE 11.2 Schematic diagram of (near-field) optical data storage system. This involves some main components: optical disk with thin coverlayer, optical pickup having objective lens, solid immersion lens (SIL) and actuator, two laser sources having different wavelength such as 405 and 780 nm and photo detector, beam splitter, polarizer, and so on.

uses a thinner substrate or cover layer and has smaller feature sizes than conventional DVDs. Such properties enable the HD-DVD system to have a large storage capacity, ranging from 18–25 GB per side. One advantage of a recording system using a thinner substrate or cover layer is that it is less sensitive to disk tilt and the capacity can be doubled by forming a double layer.

HD-DVD substrates are usually fabricated by injection-compression molding. As the optical disk substrate becomes thinner, the fabrication of disk substrates that satisfy the required optical and geometrical properties becomes more difficult, because the viscosity of the polymer melt increases due to the rapid growth of the solidified layer. During injection molding of an optical disk, the shear stress resulting from this elongation induced by the high viscosity of the polymer melt causes birefringence in the molded optical disk substrate. The solidified layer may result in many adverse effects in the geometrical and optical properties of the molded disk substrates, such as radial tilt and transcribability. It is therefore important to carefully control the growth of the solidified layer.

Figure 11.3 shows the injection molding process used in manufacturing optical disk media, which can be divided into four steps: glass mastering, mold creation, injection molding, and disk finishing. Glass mastering consists

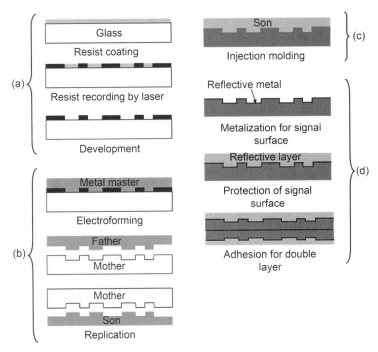

FIGURE 11.3   Manufacturing process of optical disk media using standard injection molding (a) glass mastering, (b) mold creation, (c) injection molding, and (d) disk finishing.

of the deposition of a photoresist onto glass, lithography, and development. The mold creation involves the deposition of a metal seed layer followed by an electroforming process, and then the mold for injection molding is obtained by a replication process such as hot embossing, thermal imprinting, or UV imprinting. Optical data media can then be produced by injection molding and subsequent metallization, providing a reflective surface for data read-out.

Future optical storage devices, including HD-DVD, digital versatile disk read-only memory (DVD-ROM) and digital versatile disk random access memory (DVD-RAM), will be introduced only if practical fabrication processes capable of reproducing nano- and microstructures can improve three major properties of the optical disk substrate: birefringence, tilt, and transcribability.

11.2   IMPROVEMENTS IN THE OPTICAL AND GEOMETRICAL PROPERTIES OF HD-DVD SUBSTRATES

This section provides an introduction to the effects of processing conditions on the optical, mechanical, and geometrical properties of the optical disk substrate. The mold temperature, melt temperature, packing pressure, and the pressure applied during compression all affect birefringence in injection-compression-molded polycarbonate substrates for magneto-optical disks (MODs) [10]. However, with the advent of high-density information storage devices and media that use short-wavelength lasers such as HD-DVD (which uses a 405 nm blue laser), conventional injection-molding processes are no longer adequate for producing optical disk substrates with sufficient optical, mechanical, and geometrical properties. New replication processes and improved mold designs are therefore required to produce optical disk substrates with the desired properties.

The HD-DVD-ROM and HD-DVD-RAM formats have substrates with pits and land-groove structures, and data reading by a phase change is possible. Here, a pit or land-groove structure is replicated using a mold, which is manufactured by a mastering process. High-quality replication of polycarbonate substrates in the disk-replication process is critical. Recently, injection-compression molding as a part of the disk-replication process has attracted growing attention. In the injection-compression-molding process, the molten polymer is injected into a cavity that is opened slightly, and once filling is completed, compression is used to apply a uniform pressure to the entire cavity. Using this process, optical disks can be manufactured with low distortion, uniform thickness, and reduced residual stress.

It is important to understand the effects of the flow rate of the molten polymer, packing pressure, pressure during compression, mold temperature, resin temperature, and cooling time on the optical and geometrical properties of the polycarbonate substrates in the injection-compression-molding process. It should be noted that mold temperature and compression pressure are the most important parameters.

Birefringence develops during the molding process and is an important factor in determining the optical properties of the disk. Birefringence is observed even in amorphous polymer materials when external pressure is applied or when residual stress exists in the materials. There are two types of birefringence that can develop during the molding of an optical disk: flow-induced birefringence due to polymer flow inside the mold cavity and thermally induced birefringence caused by nonuniform cooling. The birefringence in the finished disk substrate is the result of contributions from both processes, and it is difficult to separate them.

The radial tilt and the integrity of replication are regarded as important geometrical properties. Severe radial tilt is warping in the radial direction, and results in distortion of the reflected beam. Accurate replication of the pit and land-groove structures in the mold is also critical for preventing cross-talk and cross-erasing in HD-DVD systems [12]. It is important to control the injection-compression-molding process so that the integrity of the replication of the pit and the land-groove structures is maximized. It is therefore necessary to understand the effects of processing conditions and properties on the substrate. Polycarbonate DVD-RAM substrates have been fabricated by injection-compression molding, and the effects of mold temperature and compression pressure on birefringence distribution, radial tilt, and the land-groove structure have been examined experimentally [13].

In current DVD systems, birefringence of less than 100 nm, radial tilt of less than 0.4°, and a land-groove structure of 0.54 μm track pitch and 80 nm groove depth are required. This study examined the effects of processing conditions on the optical and geometrical properties of DVD substrates for effective replication of HD-DVDs. Polycarbonate substrates 120 mm in diameter and 0.6 mm thick were fabricated using injection-compression molding, as shown in Figure 11.4. They were made using optical grade polycarbonate (Lexan OQ 1020c-112), which is a common material for optical-disk substrates. The injection-compression molding machine had a clamping force of 35 t, a maximum stroke volume of 42 cm³, and a maximum injection pressure of 1690 kg/cm². The injected resin was heated to 360°C, and the temperature distribution in the cylinder was divided into four sections. From the section nearest to the sprue, the temperature distribution was 360, 380, 360, and 300°C. The filling time was 0.17 s and the packing pressure was 30–35 kgf/cm². To examine the influence of the pressure and temperature on the properties of the DVD-RAM substrates, substrates were

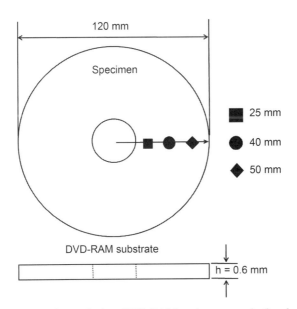

FIGURE 11.4 Dimension of the DVD-RAM substrate and the location of the measurement points.

produced with different pressures applied for 8 s after filling. The mold temperature was also varied.

Birefringence was quantified from the retardation of a reflected beam and the radial tilt was quantified from the angular deviation using a micropolariscope. The distribution of the double-path birefringence was measured using a red laser normal to the surface of the disk substrate. The location of the measurement point is shown in Figure 11.4. An atomic force microscope (AFM) was used to measure the land-groove structure in the mold and the plastic substrates, as shown in Figure 11.5. Substrates were selected at random to avoid systematic errors. The area of the region imaged using AFM was $25\,\text{m}^2$ for each measurement point. Groove depth was found to be the dominant factor affecting the integrity of replication.

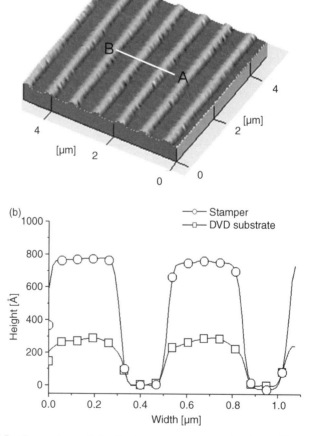

FIGURE 11.5 Comparison of the land-groove structures in the mold and the plastic substrate molded with arbitrary processing conditions.

Birefringence was the greatest near the center of a disk, and the smallest near the edges. This resulted from the velocity gradient, which in turn determined the shear rate, and decreased toward the edge of the disk. The solidified layer in the vicinity of the gate resulted in elongation of the polymer melt, causing flow-induced shear stress near the center of the disk.

As the mold temperature increased, the variation in birefringence across the disk decreased, as shown in Figure 11.6a. This was because the elevated mold temperature slowed the growth of the solidified layer during filling. Also, the elevated mold temperature prevented the development of thermal stress during compression and cooling. Birefringence increased in the vicinity of the edge as the pressure increased, as shown in Figure 11.6b. This was because the flow in the radial direction was reversed near the edge, as has been described elsewhere [1]. The birefringence was found to depend more strongly on temperature than on pressure.

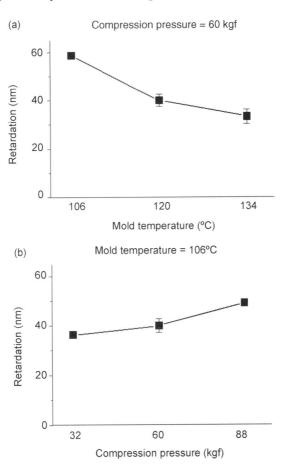

FIGURE 11.6   Effects of processing conditions on mean value of birefringence; (a) and (b).

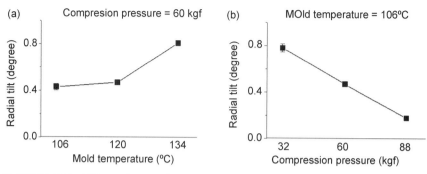

FIGURE 11.7 Effects of processing conditions on mean value of radial tilt; (a) and (b).

Higher mold temperatures increased the radial tilt, as shown in Figure 11.7a. This was because cooling was unbalanced during the early stages of mold opening. The pressure applied after filling results in uniform pressure distribution, which provides a uniform stress distribution, as explained by Santhanam and Wang [14]. Variation in the radial tilt decreased as the pressure increased, as shown in Figure 11.7b.

Increasing the mold temperature decreased the viscosity of the polymer melt and improved the integrity of the replication, as shown in Figure 11.8.

It was also found that the land-groove structure was more uniform in the radial direction as the mold temperature increased. At too high a mold temperature, however, the vitrified polymer tended to adhere to the mold, resulting in deterioration of the surface quality. Increased pressure improved the integrity of data replication and the uniformity of the land-groove structure in the radial direction, as shown in Figure 11.9. However, increased pressure also increased birefringence.

A method to prevent the formation of a solidified layer at the mold wall without increasing the mold temperature is required. Optimization of the process conditions is required to obtain high-quality disk substrates that allow accurate reproduction of nano- and microscale features.

11.3 EFFECTS OF THE INSULATION LAYER ON THE OPTICAL AND GEOMETRICAL PROPERTIES OF THE DVD MOLD

The polymer melt cools rapidly when it contacts the mold wall during filling, and a solidified layer forms at the mold wall. This increases the viscosity of the polymer melt, elongating the polymer molecules near the mold. The resulting shear stress causes birefringence in the molded optical disk substrate. Thinner substrates (such as those used in HD-DVD) have smaller effective flow channels for the polymer melt, and this causes the viscosity to

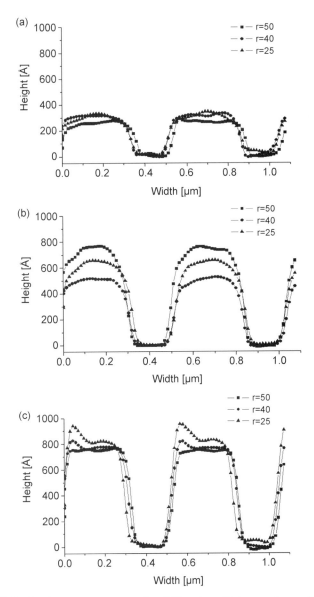

FIGURE 11.8   Land-groove structures for different mold temperatures: (a) 106°C; (b) 120°C; (c) 134°C (compression pressure $= 60 \, \text{kgf/cm}^2$).

increase. As the thickness of optical disk substrates decreases, fabrication becomes more challenging.

The solidified layer grows rapidly due to heat transfer between the polymer melt and the mold wall during filling. It is therefore very difficult to fill submicron-sized cavities with the polymer melt before it solidifies. If the

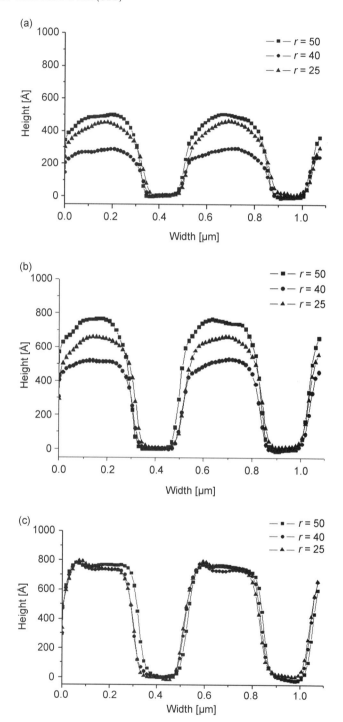

FIGURE 11.9 Land-groove structures for different compression pressures: (a) 32 kgf/cm²; (b) 60 kgf/cm²; (c) 88 kgf/cm² (mold temperature = 120°C).

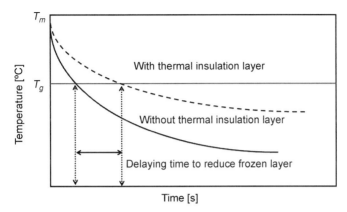

FIGURE 11.10 Cooling curve with thermal insulation layer and without thermal insulation layer.

temperature of the mold wall is maintained above the glass transition temperature, growth of the solidified layer can be slowed during filling. Maintaining the polymer melt above the glass transition temperature, T_g, provides sufficiently low viscosity and long relaxation times, which improves the properties of the disk substrate. However, ideal filling cannot be achieved because there is a limit to the temperature of the mold beyond which the melt adheres to the mold. The contact temperature is determined only by the mold temperature and the thermal properties of the mold material. A thermal insulation layer retains heat at the mold surface and slows the growth of the solidified layer causing a relaxation delay, as shown in Figure 11.10. In this way, the growth of the solidified layer can be controlled during molding. The cooling time is mainly determined by the delaying effect of the insulation layer, which is dependent on the thickness of the layer as shown in Table 11.1.

To analyze the effect of the insulation layer on the replication of DVD media, DVD-RAM disk substrates were produced with mold temperatures of 90, 100, and 110°C, and with insulation-layer thicknesses of 33 and 78 μm. The birefringence distribution and integrity of replication were measured at 25, 40, and 55 mm from the center of the disk. A comparison of radial tilt values is not valid because the insulation layer was applied to only one side, and this causes the disk substrate to warp in the radial direction, due to the

TABLE 11.1 Thermal Properties of Materials

Material	Thermal Conductivity (W/mK)	Thermal Resistance (m^2K/m)
33 μm thick polyimide	0.375	8.8×10^{-5}
78 μm thick polyimide	0.375	20.8×10^{-5}
Steel for mold	15.0	–

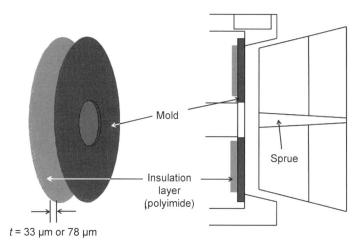

t = 33 µm or 78 µm

FIGURE 11.11 Schematic of thermal insulation layer setup.

temperature difference between the two sides of the disk. A 33 µm-thick and a 78 µm-thick polyimide layer were attached to the back side of a DVD-RAM mold, as shown in Figure 11.11.

Figure 11.12 shows a photograph of the injection-compression mold with the thermal insulation layer. Optical grade polycarbonate Lexan OQ 1020c-122 was used to mold a 120.0 mm diameter and 0.6 mm thick disk.

FIGURE 11.12 Actual injection-compression mold with a thermal insulation layer to fabricate optical disk substrate. Reprinted with permission from Ref. [15]. Copyright 2002, SPIE.

To measure the gapwise birefringence distribution, that is, the variation in the birefringence in the direction normal to the plane of the disk, specimens were cut from the disks using a diamond saw. To reduce heating due to friction, the angular velocity of the saw was maintained at between 150 and 200 rpm, and to prevent mechanical damage to the disk two Plexi-glass plates were placed one on either side during cutting. The specimens were polished using 1 μm diamond suspension solvent to minimize the effects of surface roughness on the birefringence measurements, followed by secondary polishing using 0.3 μm alumina slurry.

The land-pit structure of the disk was measured at three points before cutting, and then the spatial distribution of the birefringence was measured using a micropolariscope. Measurements were taken at five different locations in the radial direction, as shown in Figure 11.4. The gapwise birefringence distribution was measured at 30 points across the disk.

As shown in Figure 11.13, using an insulation layer the structure formed was completely filled despite the low mold temperature. The groove depth was 50 nm and the track pitch was 0.68 μm. The insulation layer had a considerable effect on the replication of surface geometry irrespective of layer thickness, and resulted in a significant reduction in the mold temperature required for complete replication.

Figure 11.14 shows the birefringence distribution in the gapwise direction of specimens cut from optical disk substrates that were fabricated without an insulation layer. In this figure, h is the half thickness of the substrate and z is the distance between the observation point and the center line. Therefore, $h/z = 1$ at the top surface of the substrate and -1 at the bottom. The birefringence distribution shows two maxima – one near the mold wall and one between the mold wall and the center. This is because the maximum shear stress occurs near the mold wall. In other words, the first peak near the mold wall results from elongation at the flow front, termed the fountain effect, and the second peak results from the growth of the solidified layer.

Birefringence depends on the velocity gradient of the melt-front, which results in shear stress. The birefringence near the center of the disk is small because the velocity gradient near the center is lower than that at the mold wall. This also indicates that the birefringence decreases toward the edge in the radial direction. This is because the radial velocity gradient of the polymer flow decreases as the flow proceeds to the outer edge.

The birefringence distribution in the radial direction when no insulating layer was used during molding is shown in Figure 11.15a, and the birefringence distribution with 33 and 78 μm thick insulation layers is shown in Figures 11.15b and c, respectively. When an insulation layer was used, the birefringence distribution changed and the peak near the mold wall disappeared. The reduced temperature difference between the mold wall and

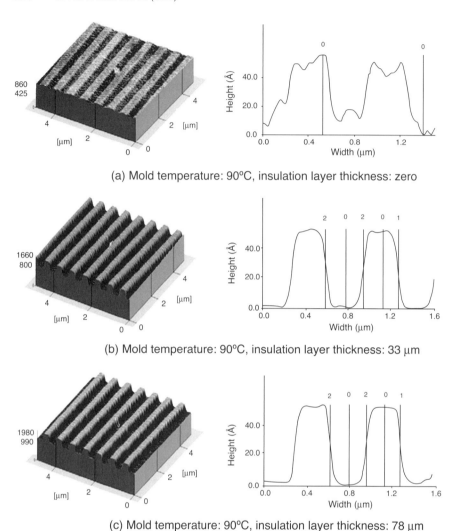

(a) Mold temperature: 90°C, insulation layer thickness: zero

(b) Mold temperature: 90°C, insulation layer thickness: 33 μm

(c) Mold temperature: 90°C, insulation layer thickness: 78 μm

FIGURE 11.13 Land-groove structure at different insulation layer thickness (a), (b), and (c). Reprinted with permission from Ref. [15]. Copyright 2002, SPIE.

the melt owing to the insulation layer decreased the degree of organization at the molecular level, and consequently the birefringence was reduced. The peak near the noninsulated mold remained, since the solidified layer is present as in the conventional process. As the thickness of the insulation layer increased, the birefringence decreased, because the thicker layer resulted in a longer delay time for thermal relaxation.

Kang et al. [16] examined the reduction in birefringence that occurs with increasing mold temperature. The decrease in the birefringence peak near

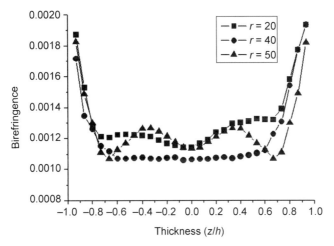

FIGURE 11.14 Gapwise distribution of birefringence without insulation layer (mold temperature: 90°C).

the insulated wall was found to be directly proportional to the increase in the mold temperature. Furthermore, an increase in the mold wall temperature resulted in an overall reduction of birefringence. The peak in the birefringence at the center in the gapwise direction also decreased with the increase in mold temperature, primarily due to a slowing of the growth of the solidified layer near the mold wall.

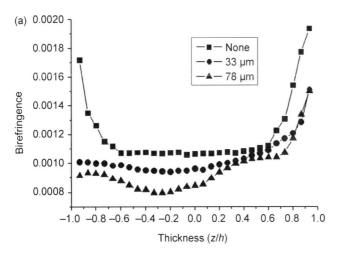

FIGURE 11.15 Gapwise distributions of birefringence for different thickness of the insulation layer: (a) $r = 25$ mm; (b) $r = 40$ mm; (c) $r = 55$ mm. Reprinted with permission from Ref. [15]. Copyright 2002, SPIE.

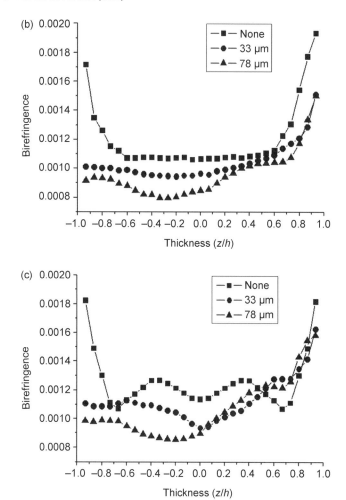

FIGURE 11.15 (*Continued*)

Figure 11.16 shows the birefringence distributions at mold temperatures of 90, 100, and 110°C when a 33 μm thick insulation layer was used, and Figure 11.17 shows the birefringence distributions at those temperatures when a 78 μm thick insulation layer was used. In both cases, the birefringence decreased as the mold temperature increased, and the reduction in the birefringence peak near the mold wall was found to be proportional to the increase in temperature.

As shown in Figure 11.16, as the mold temperature was increased from 90°C to 100°C, the birefringence near the mold wall with the 33 μm thick insulation layer was reduced, but the total birefringence was not affected. When the mold temperature was raised to 110°C, the total birefringence was

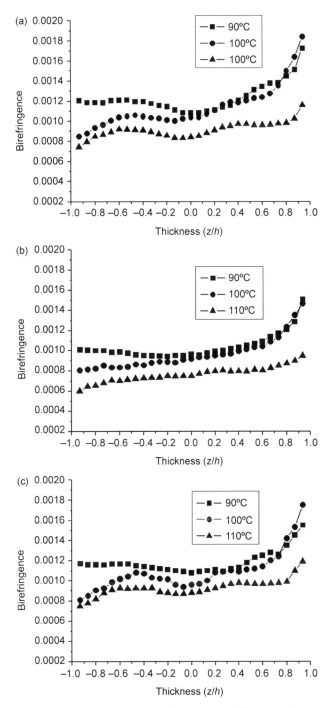

FIGURE 11.16 Gapwise distributions of birefringence for different mold temperature under the insulator layer of 33 μm: (a) $r = 25$ mm; (b) $r = 40$ mm; (c) $r = 55$ mm. Reprinted with permission from Ref. [15]. Copyright 2002, SPIE.

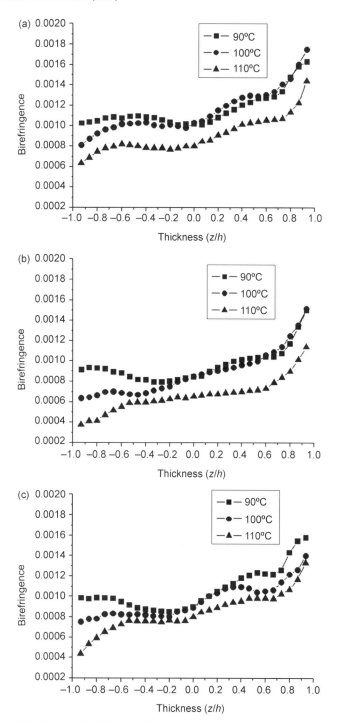

FIGURE 11.17 Gapwise distributions of birefringence for different mold temperature under the insulator layer of 78 μm: (a) $r = 25$ mm; (b) $r = 40$ mm; (c) $r = 55$ mm.

reduced, as well as the birefringence near the mold wall. It is also notable that the second peak was reduced with the increase of the mold temperature, and this is mainly because the growth of the solidified layer near the mold wall was prevented slightly due to slow cooling.

Figure 11.17 shows the birefringence distribution at mold temperatures of 90, 100, and 110°C when the 78 μm thick insulation layer was used. The total birefringence was reduced more with the thicker insulation layer. The birefringence peak near the mold wall was significantly reduced when the 78 μm thick insulation layer was used, irrespective of the mold temperature. Additionally the 78 μm thick insulation layer resulted in a reduction of the birefringence peak at the center of the disk in the gapwise direction. This is because the 78 μm thick insulation layer resulted in slower cooling of the melt.

The use of an insulation layer resulted in a significant reduction of the mold temperature required to achieve complete replication. This suggests that considerable improvements in the molding process should be possible by controlling the temperature of the mold wall.

The prediction of birefringence using more elaborate viscoelastic model is in conjunction with injection-compression molding process. Furthermore, experimental results showed that the optical and geometrical properties of optical disk substrates can be controlled and a superb optical disk substrate for HD-DVD can be fabricated by adapting the mold system to retard the growth of the solidified layer.

11.4 OPTIMIZED DESIGN OF THE REPLICATION PROCESS FOR OPTICAL DISK SUBSTRATES

With the increased application of optical and magneto-optical information storage devices with high storage density, such as CD, DVD, and MOD, optical pickup laser's wavelength is getting shorter and shorter; from the present standard of 780 nm to 405 nm Also, the track pitch is expected to be narrower than the present standard of 0.74 μm, which will greatly increase the areal density. Recently, for this submicron structured optical disk substrate production process, injection-compression molding as a part of the disk replication process is getting more attention than the simple injection molding process because the compression process produces an optical disk with small distortion, uniform thickness distribution, and reduced residual stress. However, it is difficult to determine the processing conditions for fabricating superb optical disk substrate because the properties are conflicted according to the processing condition as shown in previous sections. Currently, the determination of the processing conditions

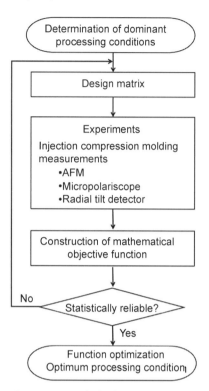

FIGURE 11.18 Scheme for process optimization

has been relied on experience or trial-and-error, which consumes much time and cost.

In this section, an efficient design methodology using the response surface method (RSM) and the central composite design (CCD) technique was developed to obtain the optimum processing conditions, which improve birefringence, radial tilt, and integrity of the replication of DVD-RAM substrate with a limited number of experiments.

Figure 11.18 shows the scheme for process optimization. Here, birefringence of less than 100 nm, radial tilt of less than 0.4°, and a land-groove structure with a 0.54 μm track pitch and 80 nm groove depth were required for high-quality recording and read-out. The response surface method (RSM) and the CCD techniques were applied to construct the so-called objective function for the optimization of results from a limited number of experiments. The analysis of variance (ANOVA) technique was used to verify the validity of the response surface, which represents the functional relationship between the processing conditions and the response values.

During injection-compression molding, the processing conditions such as flow rate of the molten polymer, packing pressure, compression pressure, mold temperature, resin temperature, and cooling time affect the optical and geometrical properties of polycarbonate substrates. As shown in the previous section, mold temperature and compression pressure were determined as dominant processing conditions. CCD was used to reduce the number of experiments. In Figure 11.19 the total sets of experimental conditions are composed of 2^k factorial design sets, four axial point sets, and one center point set as explained in Walpole and Myers [16].

The parameter k represents the number of axes or the total number of input parameters, which are the mold temperature and the pressure. Thus, a minimum of nine experimental runs, where $k = 2$ in this case, is needed to estimate the linear, quadratic, and interactive effects of the processing parameters on the integrity of the land-groove structure. For example, the working range of the processing parameters can be determined by adjusting

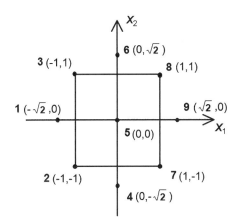

(a) Sets of coded values

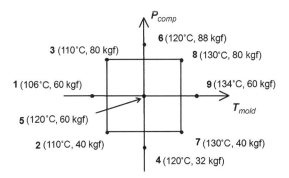

(b) Sets of actual values

FIGURE 11.19 CCD comprised the 2^k factorial design sets, four axial points sets, and one center point set. Copyright 2011, KSME.

the parameters required for producing a substrate without the land-groove structure, where $T_{mold,max} = 134°C$, $T_{mold,min} = 106°C$, $P_{comp,max} = 88$ kgf/cm², and $P_{comp,min} = 32$ kgf/cm². The upper limit of a factor was coded as $+1.4142$ and the lower limit as -1.4142 following the $2k/4$ rule, and the nine coded values obtained are shown. The actual corresponding sets of processing parameters were calculated following the nine sets of coded values. The other coded values were calculated from the following relationship:

$$x_1 = \sqrt{2} \cdot \left\{ 2T_{mold} - \left(T_{mold,max} + T_{mold,min}\right)\right\} / \left(T_{mold,max} - T_{mold,min}\right) \tag{11.1}$$

$$x_2 = \sqrt{2} \cdot \left\{ 2P_{comp} - \left(P_{comp,max} + P_{comp,min}\right)\right\} / \left(P_{comp,max} - P_{comp,min}\right) \tag{11.2}$$

where x_1 and x_2 are the required coded values of the parameters T_{mold} and P_{comp}, respectively, and T_{mold} and P_{comp} vary from $T_{mold,min}$ to $T_{mold,max}$ and from $P_{comp,min}$ to $P_{comp,max}$. To construct the response surface, nine experimental runs were made using these processing conditions.

Figure 11.20 shows the birefringence and Figure 11.21 shows the radial tilt of the substrate controlled by the design matrix. As was shown in the previous section, elevation of the mold temperature reduced the birefringence and increased the radial tilt. As the pressure increased, the birefringence increased slightly, but radial tilt improved.

Increasing the mold temperature and the pressure improved the integrity of replication, as shown in Figure 11.22. However, an excessive increase in the mold temperature resulted in the vitrified polymer adhering to the surface of the mold, as explained in section 11.1.

To establish objective functions for multiple aims means that we need to minimize the deviation, f_1, between the groove depth of the substrate and the groove depth of the mold, the birefringence, f_2, and the radial tilt, f_3. Here, f_1, f_2, and f_3 are defined as follows:

$$f_1(x_1, x_2) = \|H_{stam} - h_{sub}(x_1, x_2)\|_2 \tag{11.3}$$

$$f_2(x_1, x_2) = \|BRF(x_1, x_2)\|_2 \tag{11.4}$$

$$f_3(x_1, x_2) = \|TLT(x_1, x_2)\|_2 \tag{11.5}$$

where H_{stam}, h_{sub}, BRF, and TLT are the groove depth of the mold, the groove depth of the substrate, the birefringence, and the radial tilt, respectively. The notations for $h_{sub}(x_1, x_2)$, $BRF(x_1, x_2)$, and $TLT(x_1, x_2)$ are used to emphasize the fact that each term has functional dependency on the coded

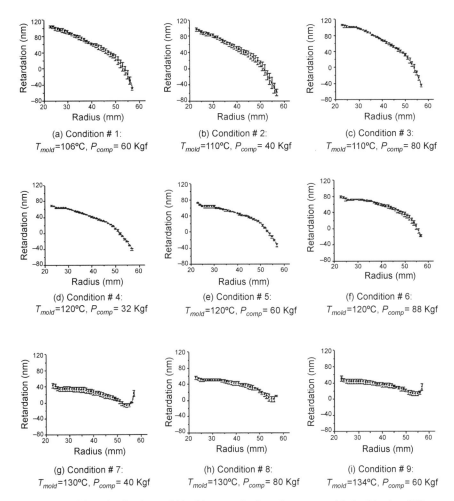

FIGURE 11.20 Distributions of birefringence in the substrates molded with nine different sets of processing condition. Copyright 2011, KSME.

values of the mold temperature (x_1) and the pressure (x_2). Since there are three independent objectives of different natures and different orders of magnitude to minimize, it is advantageous to transform each objective function into a dimensionless parameter with a magnitude between 0 and 1, such as follows:

$$f'_i(x_1, x_2) = \frac{f_i(x_1, x_2) - f_{i,\min}}{f_{i,\max} - f_{i,\min}} \quad (i = 1, 2, 3) \tag{11.6}$$

Now, the optimization problem can be recast as a determination of the coded values of the optimum mold temperature x_1 and the optimum compression

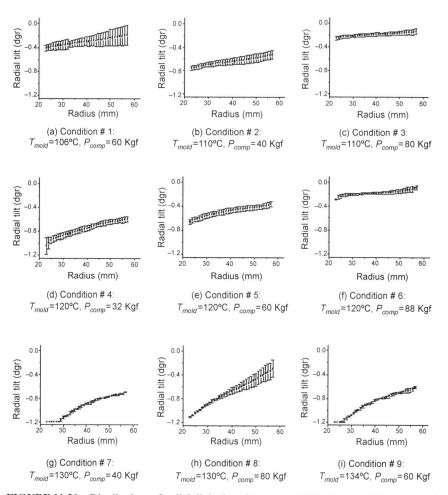

FIGURE 11.21   Distributions of radial tilt in the substrates molded with nine different sets of processing condition. Copyright 2011, KSME.

pressure x_2.

$$f(x_1^*, x_2^*) \leq f(x_1, x_2) \tag{11.7}$$

Where

$$f(x_1, x_2) = \alpha_1 f'_1 + \alpha_2 f'_2 + \alpha_3 f'_3, \ \alpha_1 + \alpha_2 + \alpha_3 = 1 \tag{11.8}$$

Here, the weights, α_1, α_2, and α_3 in Eq. (11.8) are introduced to emphasize the relative importance of each term in the objective function. From an implementation point of view, the objective function was constructed from the design matrix, which was a result of the CCD approach. Then, Eq. (11.8) was fitted using the following second-order functions in Figure 11.23 [17].

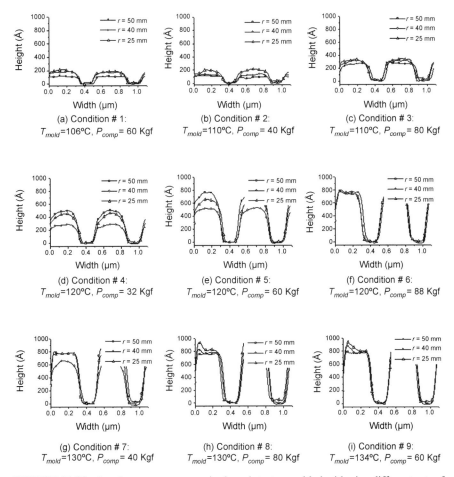

FIGURE 11.22   Land-groove structures in the substrates molded with nine different sets of processing condition. Copyright 2011, KSME.

$$f = b_0 + b_1x_1 + b_2x_2 + b_3x_1x_2 + b_4x_1^2 + b_5x_2^2 \qquad (11.9)$$

The constants b_0, b_1, b_2, b_3, b_4, and b_5 in Eq. (11.9) were obtained using the least-squares method. The adequacy of the derived function in Eq. (11.9) was tested using ANOVA for different weighting values.

The optimum processing conditions and the corresponding R^2 values calculated with various sets of weight values are shown in Table 11.2. The coefficients of determination (R^2) are above 0.90, which implies that the weight values are suitable except for case 1 ($\alpha_1 = 0.1$, $\alpha_2 = 0.5$ and $\alpha_3 = 0.4$). The obtained optimum processing conditions, clustered

TABLE 11.2 Optimum Sets of Processing Conditions for Different Weight Values

#	Weight α_1:α_2:α_3 (Hi:Bire:Skew)	R^2	Optimum Coded Values		Optimum Real Values	
			x_1	x_2	T_{mold} (°C)	P_{comp} (kgf/cm²)
1	0.1:0.5:0.4	0.87	0.32	0.59	123.2	71.8
2	0.2:0.4:0.4	0.90	0.20	0.87	122	77.4
3	0.3:0.3:0.3	0.92	0.37	0.67	123.7	73.4
4	0.4:0.3:0.3	0.93	0.42	0.61	124.2	72.2
5	0.5:0.3:0.2	0.95	0.65	0.30	126.5	66
6	0.6:0.2:0.2	0.95	0.55	0.48	125.5	69.6
7	0.7:0.2:0.1	0.95	0.59	0.43	125.9	68.6
8	0.8:0.1:0.1	0.96	0.63	0.40	126.3	68
9	0.1:0.1:0.8	0.97	−2.8	7.91	92	218.2
10	0.1:0.8:0.1	0.95	2.2	−1.58	142	28.4

around a point ($T_{mold} = 124°C$, $P_{comp} = 70$ kgf/cm²) as shown in Figure 11.24, were found to have R^2 values of approximately 0.90. These reliable optimum values were in the ranges $122°C < T_{mold} < 126.5°C$ and 66 kgf/cm² $< P_{comp} < 77.4$ kgf/cm². A series of experiments was performed using the optimum processing conditions ($T_{mold} = 124°C$, $P_{comp} = 70$ kgf/cm²), and the properties of the substrate were measured to show the validity of this methodology.

Figure 11.25 shows the measured properties under optimum processing conditions. It is noticeable that both the deviation between the groove depth of the substrate and that of the mold and the surface roughness were markedly

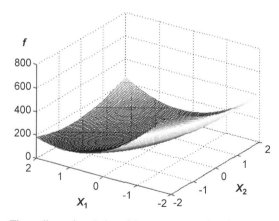

FIGURE 11.23 Three-dimensional plot of the response surface in case of $\alpha_1 = 0.3$, $\alpha_2 = 0.3$, and $\alpha_3 = 0.3$ (*f*: response surface, x_1: coded value of T_{mold}, x_2: coded value of P_{comp}).

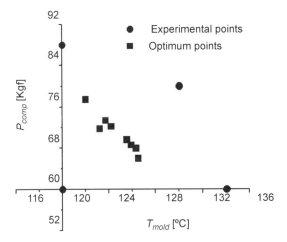

FIGURE 11.24   25 Optimum sets of processing conditions with different weight values. Copyright 2011, KSME.

reduced. Also, the birefringence and the radial tilt were reduced compared with the cases shown in Figures 11.20 and 11.21.

When polycarbonate DVD-RAM substrates were fabricated by injection-compression molding, the conflicting processing conditions were optimized. An efficient design methodology using both the CCD technique and RSM was used to obtain the optimum processing conditions, which maximized the integrity of the replication with a limited number of experiments. The optimum processing conditions that resulted in superior optical and geometrical properties were a mold temperature in the range between 122 and 126.5°C and pressure in the range between 66 and 77.4 kgf/cm². Applying this methodology to the optimization of optical and geometrical properties in optical disk substrates could have significant practical importance for mass production, saving both time and cost. Simulation of the microfluidics of the melt in the submicron structured mold cavity and active control of the mold-wall temperature (such as a heating element system) should be considered as areas for future work.

11.5 CONCLUSIONS

To fabricate high-density optical disk substrates for applications such as HD-DVD using injection-compression molding, the following methods and procedures can be considered. The mold temperature and pressure were shown to affect the birefringence distribution, radial tilt, and land-groove structure. Elevation of the mold temperature improved the integrity of the

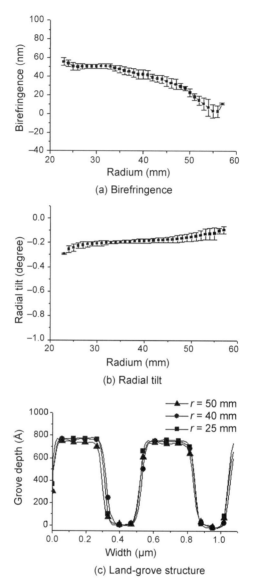

FIGURE 11.25 Measured properties in the substrate produced with the optimum processing conditions ($T_{mold} = 124°C$, $P_{comp} = 70$ kgf/cm2). Copyright 2011, KSME.

replication, decreased the overall birefringence, but increased the radial tilt. Too high a mold temperature, however, caused the vitrified polymer to stick, resulting in deterioration of the surface quality. Increasing the pressure during compression improved the integrity of replication and reduced the radial tilt, although it was found to increase the birefringence.

An insulation layer can be used to improve the viscosity of the melt. In the conventional process, there is a peak in the gapwise birefringence distribution due to solidification of the melt at the mold wall. Application of an insulation layer was found to reduce this peak and reduce the total birefringence of the disk due to the reduced temperature difference between the mold wall and the polymer melt. The insulation layer also improved the integrity of replication irrespective of layer thickness. Additionally, this suggests it may be possible to reduce birefringence by active control of the mold wall temperature.

The conflicting processing conditions were optimized using the CCD and RSM techniques to obtain optimum processing conditions, which maximized the integrity of replication with a limited number of experiments. Applying this methodology to the optimization of optical and geometrical properties in optical disk substrates could have significant practical importance for mass production.

REFERENCES

1. S. Kang, J. Kim, and H. Kim (2000) On the birefringence distribution in magneto-optical disk substrates fabricated by injection-compression molding. *Optical Engineering* 39, 689–694.

2. H. Minemura, Y. Sato, N. Tsuboi, T. Sugita, T. Fushimi, S. Yasukawa, H. Andoh, N. Muto, F. Kugiya1 and Y. Tsunoda (1993) 2.5 inch flat-type phase-change optical disk drive. *Japanese Journal of Applied Physics* 32, 5365–5370.

3. M. Shinoda, K. Saito, T. Ishimoto, T. Kondo, A. Nakaoki, N. Ide, M. Furuki, M. Takeda, Y. Akiyama, T. Shimouma, and M. Yamamoto (2005) High-density near-field optical disc recording. *Japanese Journal of Applied Physics* 44, 3537–3541.

4. D. P. Tsai and W. C. Lin (2000) Probing the near fields of the super-resolution near-fieldoptical structure. *Applied Physics Letters* 77(10) 1413–1415.

5. A. J. Bernal, M. P. Burr, G. W. Coufal, H. Guenther, H. Hoffnagle, J. A. Jefferson, C. M. Marcus, B. Macfarlane, R. M. Shelby, R. M. Sincerbox, G. T. (2000) Holographic data storage. *IBM Journal of Research Development* 44, 341–368.

6. H. E. Pudavar, M. P. Joshi, and P. N. Prasad (1999) High-density three-dimensional optical data storage in a stacked compact disk format with two-photon writing and single photon readout. *Applied Physics Letters* 74, 1338.

7. B. D. Terris, H. J. Mamin, D. Rugar, W. R. Studenmund, and G. S. Kino (1994) Near-field optical data storage using a solid immersion lens. *Applied Physics Letters* 65, 388.

8. B. W. Smith, Y. Fan, J. Zhou, N. Lafferty, and A. Estroff (2006) Evanescent wave imaging in optical lithography. *Proceedings of SPIE* 6154, 100–108.

9. K.-S. Jung, H.-M. Kim, S.-J. Lee, N.-C. Park, S.-I. Kang, Y.-P. Park (2005) Design of optical path of pickup for small form factor optical disk drive. *Microsystem Technologies* 12, 1041–1047.

10. S. Kang (2004) Replication technology for micro/nano-optical components. *Japanese Journal of Applied Physics* 43(8B) 5706–5716.

11. D. Chang, D. Yoon, M. Ro, I. Hwang, I. Park, and D. Shin (2003) Synthesis and characteristics of protective coating on thin cover layer for high density-digital versatile disc. *Japanese Journal of Applied Physics* 42, 754–758.

12. L. Satoh, S. Ohara, N. Akahira, and M. Takenaga (1998) Key technology for high-density rewritable DVD (DVD-RAM). *IEEE Transactions on Magnetics* 34(2), 337–342.

13. S. D. Moon, K. Seong, H. Kim, S. Kang, J. S. Lee, and D. Lee (2000) Improvement of optical, mechanical, and geometrical properties of DVD-RAM substrate. *5th International Symposium on Optical Storage (ISOS) (U.S.A), SPIE publication* 4085, 306–310.

14. N. Santhanam and K. K. Wang (1990) A theoretical and experimental investigation of warpage in injection molding. *SPE ANTEC Technical Papers* 36, 270–273.

15. Y. Kim, K. Seong, and S. Kang (2002) Effect of insulation layer on transcribability and birefringence distribution in optical disk substrate. *Optical Engineering* 41(9), 2276–2281.

16. R. E. Walpole, R. H. Myers, and S. L. Myers (1993) *Probability and statistics for engineers and scientists*, 5th Edition, Prentice Hall, Upper Saddle River, NJ.

17. R. O. Kuehl (2000) *Design of Experiments: Statistical Principles of Research Design and Analysis*, Duxbury Press, CA.

CHAPTER 12

Biomedical Applications

12.1 INTRODUCTION

Biosensor technologies have been intensively investigated due to their promise in life prolongation and environmental surveillance [1]. A biosensor is composed of an immobilized bioreceptor, a signal transducer, and a detector, and various types of biosensors are possible by combining such elements. Signals in biosensors can be achieved electrochemically, optically, and mechanically, and a biosensor can be classified by its transducer type as shown in Figure 12.1 [2]. Biosensors can also be classified by the use of labels such as fluorescent dye, radio active isotopes and enzyme etc. Label-free biosensors are devices that use biological or chemical receptors to detect analytes (molecules) in a sample.

Nanobio detection technology, which is a mean of acquiring large amount of bioinformation, can be a powerful tool for understanding biological phenomena at the DNA level, in the prevention of disease, and in the diagnosis for various disorders [3,4]. In recent years, biodetection systems have become much smaller and now the use of multi arrays to analyze various analytes simultaneously is common. The advantages of micro-/nanoscale biochip are as follows:

(1) Increased sensitivity
(2) Reduced reagent volumes and associated costs
(3) Increased capability for multi-agent detection
(4) Reduced processing time

For analyzing biological system on the atomic or molecular level, the introduction of nanotechnology for the fabrication of such micro-/nanobiochips is

Micro/Nano Replication: Processes and Applications, First Edition. Shinill Kang.
© 2012 John Wiley & Sons, Inc. Published 2012 by John Wiley & Sons, Inc.

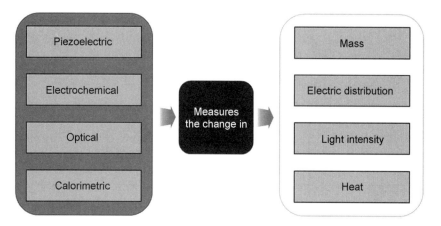

FIGURE 12.1 Classification of biosensors by transducer type.

essential. In this chapter, optical based label-free "Guided Mode Resonance" (GMR) protein sensor with nanograting structure will be introduced as an example of micro-/nanobiosensor, and also it will be explained in detail how nanoimprinting or nanoinjection molding process can be applied as tools for realizing label-free GMR protein sensor.

12.2 GMR-BASED PROTEIN SENSORS

12.2.1 Principle of GMR Protein Sensors

A biosensor incorporating a guided mode resonance protein chip is a label-free biosensor that is designed to create a sharp reflectance or transmittance peak when its nanograting structure is illuminated with white light [5]. The GMR protein chip consists of a subwavelength grating structure and high-refractive-index material deposited on the grating structure. The GMR effect arises when a subwavelength grating excites a leaky mode via evanescent-wave diffraction on a waveguide. Coupling the wave diffracted by the grating structures and guided mode in the waveguide causes reflection of the narrow bandwidth wavelengths only. Using the sharp reflection or transmission peak, the GMR protein chip measures the change in the sharp peak wavelength value (PWV) as biomolecular interactions take place on the grating surfaces [6,7]. Figure 12.2 shows a schematic diagram and the working principle of a biosensor using a GMR protein chip. Consequently, application of biochips is expected to increase rapidly with the incorporation of high-sensitivity GMR protein chips using this principle.

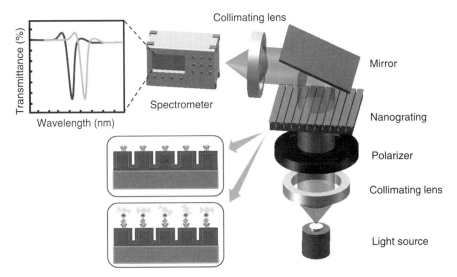

FIGURE 12.2 Schematic design and working principle of GMR protein chip.

12.2.2 Principle of Guided-Mode Resonance Effect

Guided-mode resonance or waveguide-mode resonance is a phenomenon wherein the guided modes of an optical waveguide can be excited and simultaneously extracted by the introduction of a phase-matching element such as a diffraction grating or prism. Such guided modes are also called 'leaky modes' as they do not remain guided and have been observed in one- and two-dimensional photonic crystal slabs [8].

The basic principle of GMR protein chip is coupling the diffracted wave caused by a grating structure to a guided mode in a waveguide. At resonance, almost 100% at specific wavelength of light is reflected so the detector can get a stronger signal coming back from the surface [9].

Characteristics: High-efficiency resonance, narrow spectral width, simple structure
Main Components of GMR Structures: Waveguide, grating structure
Mechanism: (1) coupling effect, (2) wave-guiding, (3) phase-matching

In general, GMR structure consists of two main components, a waveguide and a grating structure. And coupling, waveguiding, and phase-matching are mechanisms that are related to GMR protein chips [10,11].

Gratings shown in Figure 12.3a are diffractive optical elements with periods smaller than the wavelength of the incident light. Because of their small period, all higher-order diffracted waves are cut off, and only the zero-order transmitted and reflected beams propagate outside the gratings. The resonances are produced when incident light within a narrow band of wavelengths is diffracted by the grating and becomes strongly coupled to leaky modes in the waveguide region [12]. Thus, a narrow transmittance peak is generated as shown in Figure 12.3b.

Leaky Mode and Phase Analysis in Resonance Condition: The reflection resonance of the grating structure is associated with the grating

(a)

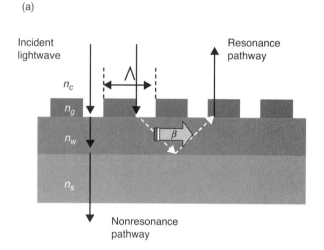

(b)

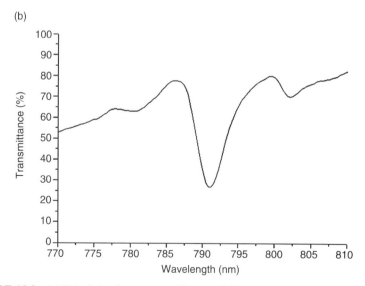

FIGURE 12.3   (a) Principle of resonance effects and (b) spectral response by resonance.

waveguide leaky modes. In a nongrating situation, the waveguide itself satisfies true-bound mode. However, a leaky mode can be achieved by adding grating layers. Leaky mode is one kind of mode that is coupled with the guided light within the waveguide with propagating light in free space. Thus, light in leaky mode can leak out of waveguide. On the other hand, an external beam can be guided in waveguide by matching leaky mode [13].

The coupling effect is a phenomenon that describes electromagnetic wave propagation going into or going out of waveguide by phase matching. To be coupled into waveguide, at least one evanescent wave should be phase-matched into waveguide mode. Phase matching means β, mode in waveguide, is corresponding to the propagation constant of diffracted order with tangential directions. $\beta wg(m)$ means mth-order propagation constant within waveguide, and k_0 ($= 2\pi/\lambda$) is the vacuum wave number. θ is the incident angle of light, i is the diffracted order and Λ means the grating period [14].

$$\beta_{wg}^{(m)} = k_0 \sin \theta_i + \frac{2\pi}{\Lambda} \qquad (12.1)$$

At resonance, all the waves diffracted out of the waveguide have a p phase shift as compared with the directly transmitted incident wave (zero-order wave). Thus, these waves will interfere destructively so that the incident wave is mostly reflected. Figure 12.4 shows the diagram of phase analysis on grating structures. And this relation is shown below [15].

Phase differences between $T_0(j)$ and $T_0(j+1)$

$$\phi = (\phi_p + 2\phi_{r(3,2)} + 2\phi_{r(3,4)} + 2\phi_d) - \phi_i = \pi \qquad (12.2)$$

($\Phi r(3,2)$ $\Phi r(3,4)$:goos-Hanchen phase shift, $\Phi d = -\pi/2$:phase shift associated with each diffraction process)

Theoretical Approaches for Nanograting Structures by RCWA: The "Rigorous coupled wave analysis" (RCWA) method is suitable for modeling and analyzing periodic grating structures. This method assumes a specific form of grating, which enables a straightforward separation of space variables. When the grating's period is of the same order of magnitude as the wavelength, rigorous methods are required to solve Maxwell's equations [16]. Fraunhofer approximations and Fourier transformation for the scalar analysis and the RCWA

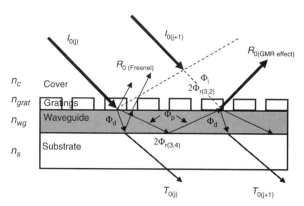

FIGURE 12.4 Phase analysis in GMR protein chips.

method for the vector analysis were employed for rectangular gratings with a period of 533 nm [17]. For this analysis, slicing a structure into multiple layers so that each layer is homogeneous in the propagation direction is required. For each layer, permittivity and EM components are represented by Fourier expansion. Figure 12.5a shows the idealized shape of the GMR protein chip. A one-dimensional-line-grating pattern was formed on a glass substrate ($n = 1.47$) and silicon nitride, which works as a high-refractive-index material was deposited on top of the nanograting. The peak spectrum bandwidth varies as the deposition thickness of the high-refractive-index materials changes. When spectral bandwidth is the sharpest, highly sensitive detection is possible. Thus, a GMR protein chip with a narrow spectral bandwidth was designed using RCWA. For better diffraction efficiency, a transverse magnetic (TM) wave was incident on the grating structures. Figure 12.5b shows the simulation result when the narrowest peak wavelength was generated by RCWA.

12.2.3 Nano Replication Process of a GMR Protein Chip for Mass Production

Like other protein chips, GMR protein chips are generally disposable, and cannot be reused because of possible errors in measurement, diagnosis reliability, and the possibility of infection. Thus, a low-cost fabrication method for commercial GMR protein chips is necessary. To achieve this, the fabrication of such protein chips using UV-imprinting and injection molding methods have been proposed. These methods are known to be outstanding methods for the mass production of polymer products and much

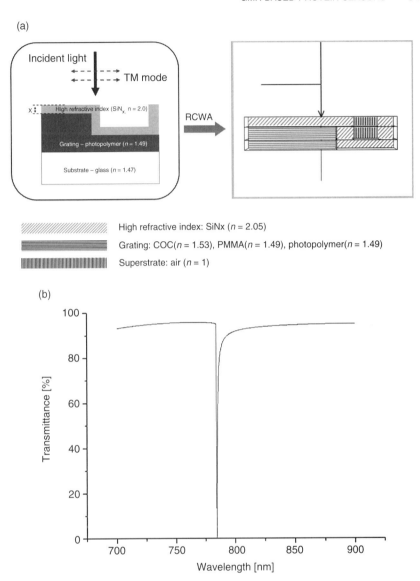

FIGURE 12.5 (a) Cross-section of GMR protein chip (b) simulation result of 70 nm deposition of high-refractive-index material.

research is being conducted on fabricating biochips through these methods [18]. In this chapter, we introduce technologies to effectively replicate nanograting patterns on a GMR protein chip by UV nanoimprinting and a nanoinjection molding process. To replicate the nanograting structure for GMR protein chip by the proposed methods, a mold with the negative shape of the designed nanograting is required. Because the widely used silicon molds are not sufficiently durable for repetitive pattern replication and especially for injection molding with high processing temperatures and pressures, a metal mold is indispensable [19]. Thus, we fabricated a metal mold using an electroforming process with a master substrate, electron beam lithography, and a UV nanoimprinting process. We investigated the effect of variables such as temperature and pressure during the UV nanoimprinting process and injection molding process to obtain high-fidelity nanograting patterns. Finally, to verify the feasibility of the fabricated protein chip, a high-refractive-index material was deposited on the UV-nanoimprinted and injection-molded nanograting pattern to produce the GMR effect, and the PWV was analyzed.

Fabrication of a GMR Protein Chip by UV-Nanoimprinting Process: A protein chip with a GMR filter was designed to have a PWV in the range 780–800 nm. Figure 12.6a shows the design specifications of the prototype GMR filter for protein chip applications. Figure 12.6b is the picture of the fabricated nanograting patterns by the UV-nanoimprinting process. The GMR filter consisted of a UV-replicated one-dimensional nanograting ($n = 1.49$) on a Pyrex glass substrate ($n = 1.47$) with a line width of 320 nm, a space width of 213 nm, a pitch of 533 nm, and a height of 160 nm with a high-refractive-index SiNx coating ($n = 2.05$) 70 nm thick. By computer simulation, we determined the sharp transmittance peak to be 784 nm with a full-width-at-half-maximum value of 0.8 nm. Transverse magnetic (TM) polarized light was projected onto the GMR filter to enhance the sharpness of the bandwidth [20].

Replication of GMR filters by UV nanoimprinting requires a mold with the negative shape of the final nanograting structure. Figure 12.7 shows an overview of the GMR filter fabrication process. Because silicon molds have durability problems in repetitive pattern replication, we fabricated a metal nanomold by electroforming [21]. To avoid deterioration of the nanostructure and master substrate, a replicated polymeric master, fabricated by UV nanoimprinting was used for the electroforming process. Nickel was selected for the electroforming material because of its properties such as hardness, durability, and adhesion. Electroforming was conducted using a precision electroforming system (Microform 500, Technotrans AG) at a process temperature of 53°C, a current of 250 A, a voltage of 4 V, and pH of 3.8.

The nanograting pattern was UV nanoimprinted on a Pyrex glass substrate using the nickel nanomold. A pressure of 31.2 kPa was applied during imprinting to minimize the inevitable shrinkage that occurs. Plasma-enhanced chemical vapor deposition was used to deposit SiNx to improve the sharpness of the PWV. Figure 12.8 shows scanning

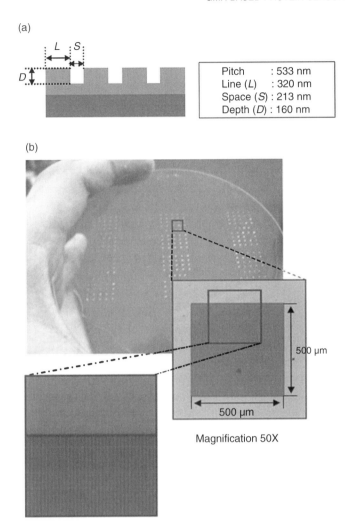

Pitch : 533 nm
Line (L) : 320 nm
Space (S) : 213 nm
Depth (D) : 160 nm

500 µm

500 µm

Magnification 50X

Magnification 2000X

FIGURE 12.6 (a) Nanograting pattern design specifications and (b) picture of fabricated nanograting patterns by UV-nanoimprinting process.

electron microscopy (SEM) images and atomic force microscopy (AFM) surface profile measurements. These indicate that the nickel mold satisfied the design specifications. However, the replicated nanograting pattern produced by UV nanoimprinting had a slight difference in pitch size and pattern height because of the polymer shrinkage effect. The biosensor with GMR protein chips has a high sensitivity when high-refractive-index material is deposited on the replicated nanograting structures. If the biosensor has sufficient sensitivity and selectivity, it will be able to detect small numbers of biomolecules in blood, improving

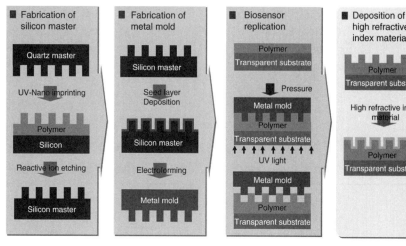

FIGURE 12.7 Fabrication processes of GMR protein chip by UV nanoimprinting.

its chances of practical use in a biosensor application. The GMR protein chips fabricated based on this research have a transmittance or reflectance peak with a narrow bandwidth, which can be a criterion of a high-resolution biosensor when material with a high refractive index is deposited. A plasma-enhanced chemical vapor deposition (PECVD) process was used to deposit the SiNx material. The temperature and pressure for the PECVD process was 250°C and 900 mTorr, respectively. Deposition thickness was determined by a computer simulation program based on RCWA. When SiNx material was deposited at 70 nm on the grating structures, the simulation result by TM polarization light showed an FWHM value of 0.8 nm, which was the narrowest and sharpest peak wavelength. PWV was created at 784 nm.

Fabrication of GMR Protein Chip by Injection Molding Process: Among various nano replication processes, the injection molding process can be the most suitable process for mass production of protein chip. For the replication of protein chip using a GMR filter with nanograting structures by injection molding process, assuring the fabrication of metal mold applied to a mold and designing optimized process parameters for fulfilling the subwavelength size cavities with polymer efficiently are important issues. Also, due to characteristics of GMR filters, fabrication of uniform pitch of grating pattern structures is mostly concerned. In this chapter, metal mold was fabricated by electron beam lithography, UV-nanoimprinting process and electroforming process to be used as a mold for an injection molding process. Protein chip using a GMR filter was replicated by injection molding as mold temperature and injection pressure changes.

A GMR protein chip composed of poly methyl methacrylate (PMMA, refractive index of $n = 1.49$) nanograting and high refractive-index SiNx (refractive index of $n = 2.05$) layer that was designed for having a PWV ranging from 790–810 nm. Figure 12.9 shows the

(a)

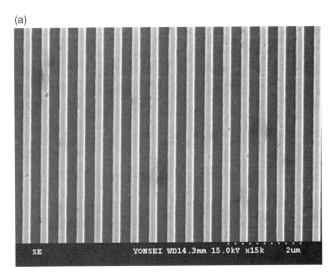

(b)

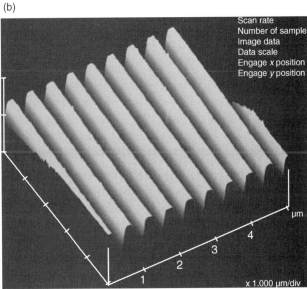

FIGURE 12.8   (a) SEM image of metal mold, (b) three-dimensional image of the metal mold by AFM. (Pitch: 538 nm, pattern height: 168 nm) (c) SEM image of the nanoimprinted polymer pattern; (d) three-dimensional surface profile image of the nanoimprinted polymer patterns by AFM measurement results (pitch: 506 nm, pattern height: 158 nm). Reprinted with permission from Ref. [22]. Copyright 2011, American Scientific Publishers.

(c)

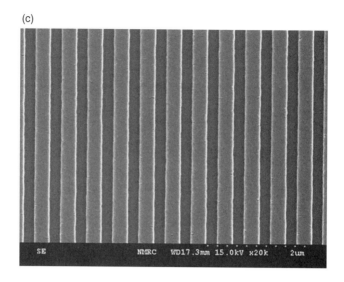

(d)

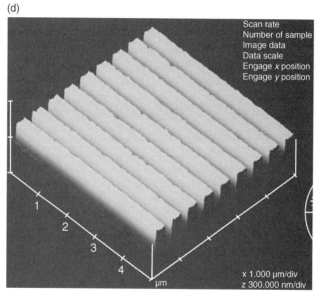

FIGURE 12.8 *(Continued)*

design specifications for this experiment. A one-dimensional nanograting with a line width of 320 nm, a space width of 213 nm, a pitch of 533 nm and a height of 160 nm were designed. 39 cells with a size of 0.5 × 0.5 mm were designed in a substrate with size of 34 × 34 mm. The distance between cells was 2.5 mm. The thickness of SiNx layer was 115 nm.

To replicate the nanograting for GMR protein chip using the injection molding process, a metal mold with nanograting structure is required.

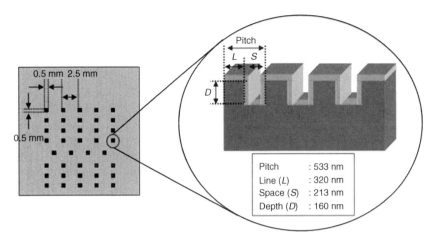

FIGURE 12.9 GMR protein chip design specifications. Reprinted with permission from Ref. [22]. Copyright 2011, American Scientific Publishers.

A metal mold using an electroforming process is used to minimize durability problems in the process of repetitive replication using injection molding. The mold fabrication procedures are illustrated in Figure 12.10. First, a quartz master with nanograting structures was fabricated using electron beam lithography. Then, a polymeric master was replicated from the original quartz master using a UV-nanoimprinting process in which a photopolymer was used as the grating material. Next, a seed layer was deposited on the polymeric master using a sputtering process to form a conductive layer for the electroforming process. Nickel was used for the seed-layer because it is sufficiently hard and stable thermally [8]. Finally a nickel mold with a size of 34×34 mm and a thickness of 1 mm were electroformed, which have a negative shape of final nanograting.

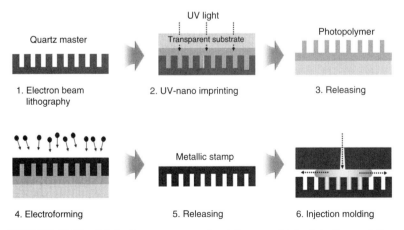

FIGURE 12.10 Fabrication processes of metal nanomold for injection molding.

(a)

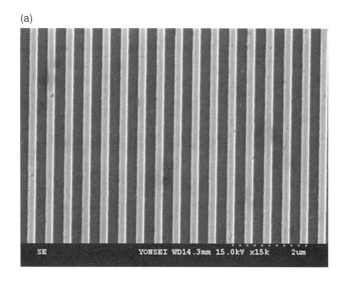

(b)

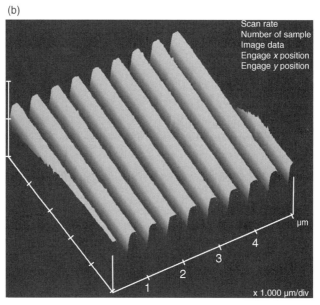

FIGURE 12.11   (a) SEM image of the metal mold; (b) three dimensional image of the metal mold by AFM (pitch: 538 nm, pattern height: 168 nm). Reprinted with permission from Ref. [22]. Copyright 2011, American Scientific Publishers.

Figure 12.11 shows a scanning electron microscope (SEM) image and atomic force microscope (AFM) image of the electroformed nickel mold. The average pitch size was 538 nm, which differed from the design value by several nanometers. The pattern height and surface profiles were

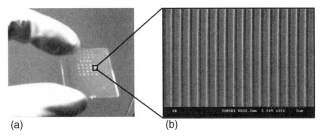

(a)   (b)

FIGURE 12.12   (a) Picture of GMR protein chip fabricated by injection molding; (b) SEM image of nanograting patterns on GMR protein chip (pitch: 506 nm, pattern height: 158 nm). Reprinted with permission from Ref. [22]. Copyright 2011, American Scientific Publishers.

verified using AFM, which showed that the pattern height of the metal mold was 168 nm.

Figure 12.12 is a picture and an SEM image of PMMA nanograting pattern fabricated using an injection molding system and the nickel mold. An injection pressure of 960 kgf/cm^2, an injection temperature of 240°C, a mold temperature of 70°C and an injection speed of 32 mm/s were selected for processing condition. We confirmed that a nanograting substrate was fabricated without any deformation when injection molding proceeded at a mold temperature of 70°C. The SEM image in Figure 12.12b showed that the pattern was transferred excellently. In the nanoinjection molding process, the mold temperature and injection pressure, time, and speed play important roles in the nanomolding process. To examine the effects of processing pressure on the replication quality of nanograting and the uniformity of injection-molded parts, the surface profiles of fabricated nanograting were measured by AFM. Figure 12.13a shows the three-dimensional surface profile image obtained from AFM measurement results. Figure 12.13c shows the comparison of surface profiles of injection-molded nanograting fabricated at a pressure of 840, 960, 1080, and 1200 kgf/cm^2 and that of nickel mold. The mold temperature, injection temperature, and speed were fixed at 70°C, 240°C and 32 mm/s, respectively. It was noted that the pattern height was regular, without huge changes at injection pressures above 960 kgf/cm^2. At a pressure of 960 kgf/cm^2, the pattern height was 158 nm and it was close to the design value of 160 nm. To confirm the uniformity of the nanograting by the injection molding process, surface profiles of the grating pattern at five points on the same nanograting were analyzed (Figure 12.13b). Figure 12.13d shows the comparison of surface profiles at five different points of a sample fabricated at a pressure of 960 kgf/cm^2. As Figure 12.13d shows, each pattern was fabricated within error ranges comparable to the design value.

(a)

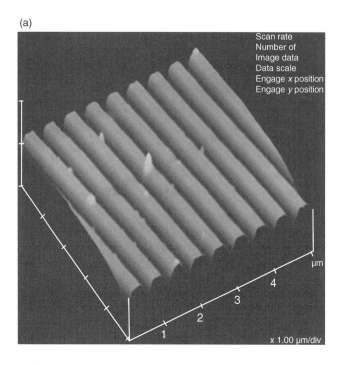

(b)

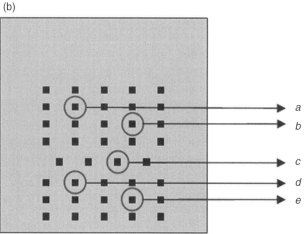

FIGURE 12.13 (a) Three-dimensional surface profile image of the injection-molded nanograting patterns obtained from AFM measurement results; (b) location of sampling points for uniformity test; (c) comparison of surface profiles of injection-molded nanograting fabricated at different pressures and that of nickel mold; (d) comparison of surface profiles of injection-molded nanograting at five different sampling points. Reprinted with permission from Ref. [22]. Copyright 2011, American Scientific Publishers.

(c)

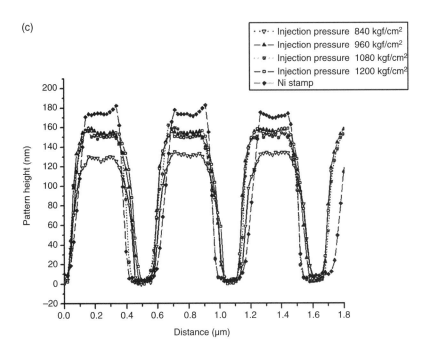

(d)

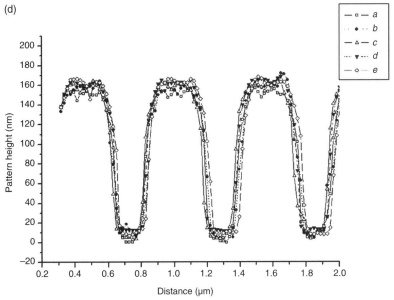

FIGURE 12.13   (*Continued*)

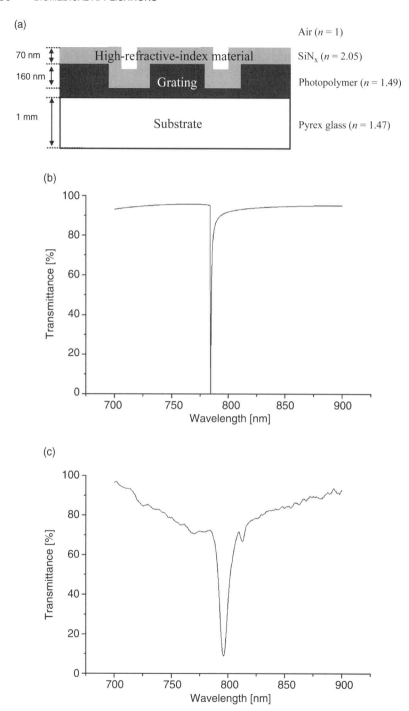

FIGURE 12.14 (a) Simulation model of UV-nanoimprinted GMR protein chip; (b) simulated transmittance spectrum result; (c) PWV result by white light system. Reprinted with permission from Ref. [22]. Copyright 2011, American Scientific Publishers.

12.2.4 Feasibility Test of GMR Protein Chip

To generate a transmittance or reflectance peak with a narrow bandwidth by GMR effect, which can be a criterion of a high-resolution biosensor, a high-refractive-index material should be deposited on the replicated nanograting structure. PECVD can be used to deposit SiNx at a temperature of 80°C, a pressure of 900 mTorr, and a deposition speed of 11 nm/min. The deposition thickness was determined by a computer simulation program based on RCWA to obtain the narrowest and sharpest peak profile. Figure 12.14a shows the simulation model of the UV-nanoimprinted GMR protein chip, consisting of a nanograting and a 70 nm thick SiNx layer. Because of the high step coverage of the PECVD process, the deposited layer on the side wall was also considered in the simulation model. Figure 12.14b shows the simulated transmittance spectrum of the GMR protein chip having a SiNx layer

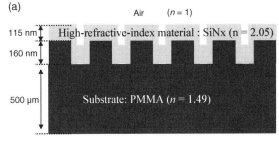

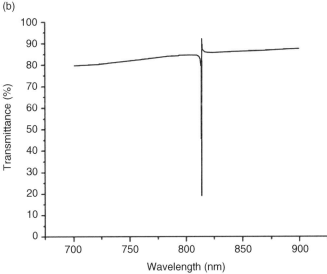

FIGURE 12.15 (a) Simulation model of injection-molded GMR protein chip; (b) simulated transmittance spectrum result; (c) PWV result by white light system. Reprinted with permission from Ref. [22]. Copyright 2011, American Scientific Publishers.

(c)

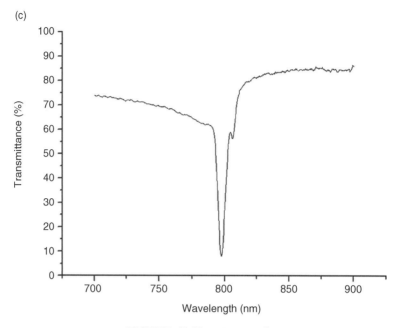

FIGURE 12.15 (*Continued*)

thickness of 70 nm, with a TM polarization light (polarization direction is incident.) The PWV was at 784 nm and the FWHM of the peak was 0.8 nm. Figure 12.14c shows the result of the PWV measurement for a GMR protein chip with a grating pitch of 506 nm and a SiNx layer thickness of 70 nm fabricated by injection molding and the PECVD process. PWV was at 796 nm and the FWHM was 8.6 nm.

Figure 12.15 shows the comparison of peak wavelength value shift by simulation and experimental results using an injection-molded GMR protein

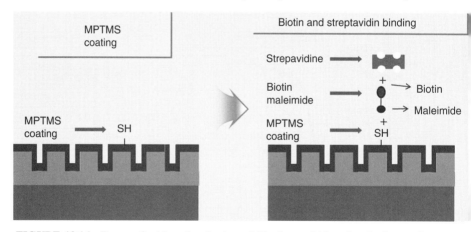

FIGURE 12.16 Process for biomolecules immobilization and biomolecular interactions.

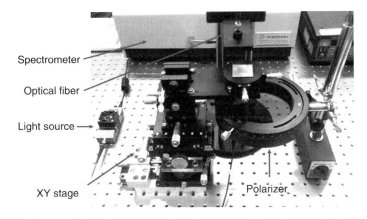

FIGURE 12.17   System setup for a detection of biomolecules binding.

chip. Figure 12.15a shows the simulation model of the injection-molded GMR protein chip, consisting of a nanograting and a 115 nm thick SiNx layer. Figure 12.15b is a result of a transmission spectrum when TM polarized light was incident on the GMR protein chip. The peak wavelength value was at 814 nm and the FWHM was 0.2 nm. The measured values were 797.6 nm PWV and 6.9 nm FWHM each. Reasons for these differences may include

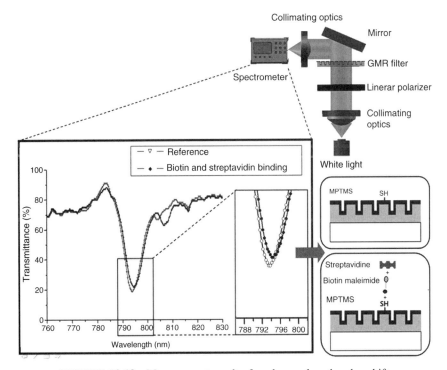

FIGURE 12.18   Measurement result of peak wavelength value shift.

changes in the grating pattern shape arising from polymer shrinkage effects during injection molding of the nanograting patterns using injection molding. Pattern defects during the replication process could be another reason for the deteriorating GMR effect. However, investigating the PWV shift with biomolecular interactions, we demonstrated the possible use of GMR protein chip as a protein sensor.

Peak wavelength value shift was confirmed through biomolecular interactions with the fabricated GMR protein chip. A UV-nanoimprinted GMR protein chip was used for the test and biotin and streptavidin were used as the detecting material to reduce nonspecific binding effects. The process is given in Figure 12.16. Figure 12.17 shows the system setup of a GMR protein sensor for the detection of biomolecules binding. The peak wavelength value shifted from 793.81 to 794.45 nm as the thickness and refractive index on top of the grating changed shown in Figure 12.18. From the result, we can conclude that the proposed method can be used for fabricating a GMR protein chip suitable for use as a biosensor.

12.3   CONCLUSIONS

UV nanoimprinting and injection molding methods, which are the most promising processes for commercial mass production of polymeric structure, were proposed to fabricate nanograting for GMR protein chip. Polymeric master nanopatterns that contain nanograting structures were fabricated using electron beam lithography and UV nanoimprinting. The electroforming process was used to make a metal mold and then UV nanoimprinting and nanoinjection molding were performed to replicate nanograting structures for GMR protein chip. The replication quality and uniformity of UV-nanoimprinted and injection-molded nanograting were analyzed using SEM and AFM measurements, and the process parameters for each fabrication methods were optimized. Finally a high-refractive-index layer was deposited on the UV-nanoimprinted and injection-molded nanograting and the simulated and measured transmission spectra of GMR protein chip were compared. To examine the feasibility of the proposed process, experiment on peak wavelength value shift by biotin and streptavidin binding effect was performed.

REFERENCES

1. D. J. Claremont (1987) Biosensors: clinical requirements and scientific promise. *Journal of Medical Engineering & Technology* 11, 51–56.

2. J. P. Chambers, B. P. Arulanandam, L. L. Matta, A. Weis, and J. J. Valdes (2008) Biosensor recognition elements. *Current Issues in MolecularBiology* 10, 1–12.

3. H. Kim, J. Doh, D. J. Irvine, R. E. Cohen, and P. T. Hammond (2004) Large area two-dimensional B cell arrays for sensing and cell-sorting applications. *Biomacromolecules* 5, 822–827.

4. T. Kubik, K. Bogunia-Kubik, and M. Sugisaka (2005) Nanotechnology on duty in medical applications. *Current Pharmaceutical Biotechnology* 6, 17–33.

5. D. L. Brundrett, E. N. Glytsis, and T. K. Gaylord (1998) Normal-incidence guided-mode resonant grating filters: design and experimental demonstration. *Optics Letters* 23, 700–702.

6. L. D. Block, Patrick. C. Mathias, N. Ganesh, S. I. Jones, B. R. Dorvel, V. Chaudhery, L. O. Vodkin, R. Bashir, and B. T. Cunningham (2009) A detection instrument for enhanced-fluorescence and label-free imaging on photonic crystal surfaces. *Applied Physics Letters* 17, 13222–13225.

7. B. Cunningham, J. Qiu, P. Li, and B. Lin (2002) Enhancing the surface sensitivity of colorimetric resonant optical biosensors. *Sensors and Actuators B: Chemical* 87, 365–370.

8. A. N. Reshetilov and A. M. Bezborodov (2008) Nanobiotechnology and biosensor research. *Applied Biochemistry and Microbiology* 44, 1–5.

9. S. S. Wang and R. Magnusson (1993) Theory and applications of guided-mode resonance filters. *Applied Optics* 32, 2606–2613.

10. S. Tibuleac and R. Magnusson (1997) Reflection and transmission guided-mode resonance filters. *Optical Society of America* 14, 1617–1626.

11. M. G. Moharam and T. K. Gaylord (1981) Rigorous coupled-wave analysis of planar-grating diffraction. *Journal of Optical Society of America* 71, 811–818.

12. M. Flury, A. V. Tishchenko, and O. Parriaux (2007) The leaky mode resonance condition ensures 100% diffraction efficiency of mirror based resonant gratings. *Journal of Lightwave Technology* 25, 1870–1878.

13. S. Nilsen-Hofseth and V. Romero-Rochin (2001) Dispersion relation of guided-mode resonances and Bragg peaks in dielectric diffraction gratings. *The American Physical Society* 64, 036614–036623.

14. T. K. Gaylord and M. G. Moharam (1982) *Planar Dielectric Grating Diffraction Theories. Applied Physics B* 28, 1–14.

15. J. Hong, K. H. Kim, J. H. Shin, C. Huh, and G. Y. Sung (2007) Prediction of the limit of detection of an optical resonant reflection biosensor. *Optics Express* 15, 8972–8978.

16. N. P. van der Aa1 (2006) Diffraction grating theory with RCWA or the C method. *Mathmatics in Industry* 8, 99–103.

17. C. Lee, K. Hane, and S. Lee (2008) The optimization of sawtooth gratings using RCWA and its fabrication on a slanted silicon substrate by fast atom beam etching. *Journal of Micromechanics and Microengineering* 18, 045014–045021.

18. D. Kim, J. Kim, Y. Ko, J. Kim, K. Yoon, and C. Hwang (2008) Experimental characterization of transcription properties of microchannel geometry fabricated by injection molding based on Taguchi method. *Microsystems Technology* 14, 1581–1588.

19. H. Kim, S. Shin, J. Han, J. Han, and S. Kang (2009) Fabrication of metallic nanostamp to replicate patterned substrate using electron-beam recording. *Nanoimprinting and Electroforming, IEEE Transactions on Magnetics* 45, 2304–2307.

20. S. S. Wang and R. Magnusson (1993) Theory and applications of guided-mode resonance filters. *Applied Optics* 32, 081103–081105.

21. B. Cunningham, P. Li, S. Schult, B. Lin, C. Baird, J. Gerstenmaier, C. Genick, F. Wang, E. Fine, and L. Laing (2004) Label-free assays on the BIND system. *Journal of Biomolecular Screening* 9, 481–490.

22. E. Cho, B. Kim, S. Choi, J. Han, J. Jin, J. Han, J. Lim, Y. Heo, S. Kim, G. Y. Sung, and S. Kang (2011) Design and fabrication of label-free biochip using a guided mode resonance filter with nanograting structures by injection molding process. *Journal of Nanoscience and Nanotechnology* 7(1), 417–421.

Micro/Nano Replication: Processes and Applications, First Edition. Shinill Kang.
© 2012 John Wiley & Sons, Inc. Published 2012 by John Wiley & Sons, Inc.